COMMUNITY GEOGRAPHY

GIS in Action

LYN MALONE • ANITA M. PALMER • CHRISTINE L. VOIGT

ESRI PRESS

REDLANDS, CALIFORNIA

ESRI
  Community Geography: GIS in Action Teacher's Guide
  ISBN 1-58948-051-1

First printing June 2003.

Printed in the United States of America.

Library of Congress Cataloging-in-Publication Data
Malone, Lyn.
   Community geography : GIS in action teacher's guide / Lyn Malone,
Anita M. Palmer, Christine Voigt.
    p. cm.
  Includes bibliographical references.
  ISBN 1-58948-051-1 (pbk. : alk. paper)
  1. Geographic information systems. I. Palmer, Anita M.
II. Voigt, Christine (Christine L.), 1970– III. Title.
G70.212.M2833 2003
910'.285—dc21   2003009537

Published by ESRI, 380 New York Street, Redlands, California 92373-8100.

Books from ESRI Press are available to resellers worldwide through Independent Publishers Group (IPG). For information on volume discounts, or to place an order, call IPG at 1-800-888-4741 in the United States, or at 312-337-0747 outside the United States.

With affection and deep appreciation we dedicate this book

to Charlie Fitzpatrick, George Dailey, and Angela Lee,

whose indefatigable enthusiasm never fails to inspire us.

Acknowledgments

This book would have been far more difficult to write and significantly less complete without the interest, advice, and assistance of many people. For their support and generosity we would like to thank the following:

The K–12 team at ESRI: Charlie Fitzpatrick, George Dailey, and Angela Lee. Their unwavering commitment to advancing the educational application of GIS has fueled this project from the beginning. Without their unflagging and energetic support this book would never have become a reality.

At ESRI in Redlands: The authors of *Community Geography: GIS in Action,* Kim Zanelli English and Laura Feaster, provided inestimable assistance throughout our writing process. Their meticulous attention to detail and thoughtful insights were critical in both the overall book design and in the editing of lesson content as well. They coordinated the efforts of the writing team and kept us on task and on track throughout the project. Additionally, we would like to thank Claudia Naber, whose editorial expertise has provided consistency and continuity to the final work. Finally, we would like to thank ESRI® instructor Jamie Parrish for providing the Human Geocoder activity in module 2.

We would also like to acknowledge the National Geographic Society for supporting this project by granting us permission to incorporate the language of the National Geography Standards, *Geography for Life,* within these lessons.

Four individuals deserve special recognition for their careful and thorough reviews of the book's materials before they went to press: Dr. Joseph Kerski, geographer with the United States Geological Survey; Gerry Bell, of Port Colborne High School in Port Colborne, Ontario; Dr. Al Lewandowski, of Port Huron Middle School, Port Huron, Michigan; and Angela Lee, with the ESRI K–12 Schools and Libraries Program. Dr. Kerski, Mr. Bell, and Dr. Lewandowski are veteran educators with extensive experience using GIS in the classroom. Angela Lee is an experienced developer of educational applications for ArcView® software. Their comments and suggestions were invaluable in the final editing process.

We would also like to take this opportunity to acknowledge a number of individuals whose friendship and ideas have shaped, supported, and enhanced our own development as both educators and GIS activists. Indirectly, they have each had a significant hand in propelling us down the path to this book. The first in this group is Gil Grosvenor. His commitment to geography education and his generous support of its teachers through the National Geographic Society's Alliance network transformed our classrooms and redirected our lives. We would also like to say a special thank you to Kate Collins Dailey for bringing the three of us together at the National GIS Institute in 1998. Kate has been a tireless advocate for geography education in general and GIS-enhanced education in particular. Similarly, we would like to thank all of the GIS "Instithooters" and K–12 ATP colleagues who have shared their time, their ideas, and sometimes even their students with us since 1998. Without that network of fellow travelers, we would not have reached this point in the road ourselves. Last, our heartfelt thanks to our husbands and families whose understanding and patience gave us the freedom to follow our dreams. Roger Palmer deserves special commendation for his unvarying and cheerful willingness to lend his expertise at critical moments.

Foreword

I'm convinced that once students learn more about geography it will appeal to their natural curiosity about the world and they will find the discipline to be exciting and fun, especially with the aid of the stimulating technologies available today. Last year the results of the latest National Assessment of Educational Progress (NAEP) report for geography were released by the Department of Education. They clearly demonstrated that kids who had access to more and better technology did better on the test. In other words, kids in schools that have more technology learn more geography, it is that simple.

In 1985, when *National Geographic* decided to turn the spotlight on geography education we made it clear that we would work directly with teachers. In fact, the very first thing we did was bring teachers to our headquarters in Washington for a summer geography institute and we did that year after year until we built a core of teachers from across the country. These teachers went back to work with geography professors and other teachers in their respective states through our geography alliance network. They were then and still are the leading edge of geography education reform in the United States.

So it is very appropriate that *Community Geography: GIS in Action Teacher's Guide* is written by three of these teachers and I highly commend them for the effort. They have created an inspiring book with hands-on, teacher-friendly exercises to build GIS skills. They also suggest practical tips for undertaking GIS projects.

This is how we are going to get high-quality, meaningful geography back into schools across the United States. And, as we were recently reminded by a National Geographic-Roper study, we still have a lot of work to do.

Despite the devastating 9/11 attacks on the United States and the subsequent media spotlight on the Middle East and Central Asia, 83 percent of young Americans aged 18 to 24 could not find Afghanistan on a world map, according to a National Geographic-Roper study on geographic literacy among young adults that we announced last November. More young Americans in the study knew that the island featured in last season's TV show "Survivor" is in the South Pacific than could find Israel.

Young people in the other predominantly English-speaking countries surveyed—Canada and Great Britain—fared almost as poorly overall as those in the United States. And none of the nine countries surveyed earned an excellent mark. Worldwide, three in ten couldn't find the Pacific Ocean, which covers 33 percent of the earth. In no nation could even half of young adults surveyed locate Israel on a map of the Middle East and Asia. Young adults worldwide also showed scant knowledge of world population issues or geography in the context of nuclear weapons.

The Geographic Literacy Study was conducted by RoperASW in June and July 2002. The survey polled more than 3,000 young adults, ages 18 to 24, in Canada, France, Germany, Great Britain, Italy, Japan, Mexico, Sweden, and the United States. Top scorers were Sweden, Germany, and Italy. Mexico ranked last. Americans, who came in next to last, expressed an exaggerated image of America's size—fully 30 percent estimated the U.S. population to be a billion or more. The correct response in the survey was 150–350 million.

The study found that young Americans were the least likely among their counterparts to know that Afghanistan is where the Taliban and al Qaeda were based. Less than half the Americans could identify France, the United Kingdom, or Japan on a world map. Fewer than two in three could find China on a map of the Middle East/Asia, and more than half—56 percent—were unable to locate India, home to 17 percent of people on earth. Just half of young Americans could find New York, one of the nation's most populous states.

This of course is appalling. If young Americans can't find places on a map, or "get" the relationship between where they are and where others are, how can they possibly understand current events, other cultures, or economic and natural-resource issues with which their generation will struggle?

We at National Geographic remain committed to making sure the next generation of young adults is better prepared to be informed citizens and responsible stewards of the planet. But we can't do this alone. I can't tell you enough how much we appreciate teachers like Anita, Lyn, and Christine and the thousands of other teachers across the country like them who are at the very cutting edge in bringing geography and these new technologies into the classroom in a useful and teacher-friendly way.

As Margaret Mead once said, "Never doubt that a small group of thoughtful committed people can change the world: indeed it's the only thing that ever has!"

Gilbert M. Grosvenor
Chairman, National Geographic Society
April 8, 2003

How to use this book

This guide was written for teachers who want to use *Community Geography: GIS in Action* with their own students. *Community Geography: GIS in Action Teacher's Guide* provides teachers with all the resources they will need to complete the GIS exercises and "on your own" projects with students. These resources include lesson plans; correlations to national geography, science, and technology standards; assessments; evaluation rubrics; and extensive teacher tips on completing community GIS projects.

The authors are all experienced classroom teachers who have tried to anticipate needs, issues, and questions that might arise when working with the book's exercises and implementing its projects with a group of students.

Where to begin

First, you need to make sure your school has ArcView 3.x software and a copy of *Community Geography: GIS in Action*. Refer to the *Community Geography: GIS in Action* appendixes for details on software and computer operating system requirements. Teachers and students with a basic introductory knowledge of GIS should begin with the module 1 exercise, "Explore and label community features data for a city visitors map." This basic exercise and assessment will refresh your students' GIS skills and introduce them to the use of community-scale data. After you have completed module 1 with your students, you are ready to choose a module topic that interests your students or is relevant to your curriculum.

Start each module by having your students read the case study, discuss it, and then complete the exercise and assessment. Depending on your class schedule, you may have time to adapt part or all of this GIS project in your own community. If you do, be sure to refer to the teacher tips in the module's "on your own" section for suggestions on how to manage the project with your students.

Each module provides you with the following essential elements:

Lesson plan for the case study and exercise

- *Overview:* A short summary gives you a snapshot of the lesson.
- *Estimated time and materials:* Time and materials for each lesson are included for your planning.
- *Standards and objectives:* A list of the National Geography Standards (as published in *Geography for Life: The National Geography Standards,* 1994) covered in this lesson for middle school and high school students is included here. Specific learning objectives give you more detailed information about the content of each lesson.
- *GIS skills and tools:* Important GIS skills and tools that are used in this lesson are summarized in list form and also by the five steps of the geographic inquiry process.
- *Teacher notes:* Teacher tips on how to introduce each lesson, technical aspects of the exercise, and how to conclude and process the lesson with students. Includes reproducible student handouts or transparencies.
- *Assessments:* Reproducible assessment handouts for middle school and high school students require students to demonstrate their understanding of concepts and GIS skills learned in the exercise.
- *Rubrics:* Rubrics for the middle school and high school assessments allow teachers to evaluate how well each student fulfilled the academic standards of the lesson.
- *Answer keys:* These keys provide answers for questions asked in the GIS exercise. They are located together at the end of the book.

"On your own" project guidelines

This section includes teacher-specific guidelines and tips for implementing a similar project in your own community. The section includes the following components:

- *Overview:* A short summary gives you an outline of the community GIS project.
- *Estimated time and materials:* Time and materials for each community GIS project are included for your planning.
- *Standards and objectives:* A list of the National Geography Standards (as published in *Geography for Life: The National Geography Standards,* 1994) covered in this project for middle school and high school students is included here. Specific learning objectives give you more detailed information about each project.
- *GIS skills and tools:* Important GIS skills and tools that are used in this lesson are summarized in list form and also by the five steps of the geographic inquiry process.
- *Teacher notes:* Tips for teachers on each step of the geographic inquiry process are included here.
  - *Ask a geographic question:* Suggestions for how to develop appropriate geographic questions given your student population and time available. Tips on how to narrow your project focus are included with each module.
  - *Acquire geographic resources:* Specific teacher tips on how to collect field data with large groups of students and how to work with local GIS users are some of the points explored in this section.
  - *Explore geographic data:* A summary of technical data issues and suggestions for preliminary data exploration.
  - *Analyze geographic information:* Various analysis techniques and suggestions are described for a variety of geographic questions.
  - *Act on geographic knowledge:* For each community GIS project, there is a list of ways your students can share their results with others.

In addition to the valuable material included with each module, there is a comprehensive references and resources section at the back of *Community Geography: GIS in Action Teacher's Guide,* and additional valuable information at the book's Web site *(www.esri.com/communitygeography).*

Now that you have a basic understanding of how this book can help you use GIS with students to make a difference in your community, you are ready to begin. Start with the module 1 GIS warm-up exercise, delve into a case study of interest, or begin planning your own community GIS project.

Contents

Dedication   v
Acknowledgments   vii
Foreword   ix
How to use this book   xi
Introduction   xv

Module 1   **GIS basics**
Lesson   1

Module 2   **Reducing crime**
Lesson   11
On your own   23

Module 3   **A war on weeds**
Lesson   27
On your own   37

Module 4   **Tracking water quality**
Lesson   43
On your own   52

Module 5   **Investigating point-source pollution**
Lesson   57
On your own   66

Module 6   **Getting kids to school**
Lesson   71
On your own   80

Module 7   **Protecting the community forest**
Lesson   83
On your own   93

Module 8   **Selecting the right location**
Lesson   99
On your own   108

References and resources   113
Correlation to National Geography Standards   117
Correlation to National Science and Technology Standards   119
Answer keys   121

Introduction

The familiar phrase "think globally, act locally" reflects a fundamental concept embraced by skilled teachers in every discipline. We want our students to be aware of the "big picture"—the interconnected network of global challenges that confront our planet in the twenty-first century. At the same time, our classroom experience has taught us that the best way to achieve this goal is by exploring the expression of these complex issues in our students' own world. Any topic—whether it be the overuse of nonrenewable resources, the spread of disease, or poverty—becomes much more compelling when, instead of being a distant abstraction, it is a real factor affecting the students' day-to-day lives. *Community Geography: GIS in Action Teacher's Guide,* written by teachers for teachers, provides detailed instructions for using the power of GIS technology to bring this principle to life in your own classroom.

This book is intended to be a teacher's companion guide for *Community Geography: GIS in Action.* Because we are all teachers, we have written this guide from a teacher's perspective. In it we provide practical and pedagogical recommendations for the successful use of *Community Geography: GIS in Action* in a classroom environment. The guide includes a comprehensive set of tips, tools, and guidelines to help you implement the exercises and projects described in the companion volume: how to identify an issue for study, how to locate and prepare appropriate data, how to use the many analytical functions of GIS to explore that data, and how to identify meaningful ways to act upon and share the conclusions that you reach. For each module, we have included lesson plans, student handouts, assessments, rubrics, and detailed recommendations and advice to assist you in designing and carrying out similar projects of your own. The valuable information included in this guide will support you and your students as you apply GIS and the methods of geographic inquiry to an issue in your community.

Not only is this a practical, how-to guide, but it is a philosophical guide for expanding the concept of "education" in your classroom. The projects in *Community Geography: GIS in Action Teacher's Guide* apply and integrate standards-based academic content with real-world community issues. In these projects, walls separating the traditional classroom from its surrounding community fall away as students develop local partnerships to gather data, analyze patterns, and develop solutions. Students come away from the experience having acquired not only key academic content, but essential problem-solving skills and practical career skills as well. They have experienced the process and pride of responsible citizenship. Additionally, the learning reflected in these case studies is based on procedures endorsed by current educational reform movements; it is hands-on, interdisciplinary, and problem-based.

Community Geography: GIS in Action is a ground-breaking resource that illustrates how the power of GIS can be harnessed to foster the practices of responsible citizenship and to demonstrate that the actions of one person truly can make a positive contribution to the community and, ultimately, to the world. *Community Geography: GIS in Action Teacher's Guide* will give teachers the necessary tools to channel that power to their own students. By addressing global issues at home, those students can prepare for their roles as the decision makers of tomorrow and help make our planet a safer and more sustainable world today.

Lyn Malone, Anita M. Palmer, and Christine L. Voigt

Module 1: GIS basics

Lesson

Exercise: Explore and label community features data for a city visitors map

Overview

This exercise reviews the basic concepts and skills of GIS that will prepare your students for the lessons ahead. Students will create a map for history teachers who are visiting the Lincoln Memorial Shrine in Redlands, California. They will start ArcView, navigate to the prepared project and data, and use a variety of ArcView tools and skills to enhance the map.

We recommend that students spend one class period on the step-by-step exercise and one class period on the assessment to gain a functional knowledge of GIS and ArcView skills.

Estimated time

45-MINUTE CLASS	ACTIVITY
1	Introduction and exercise
2	Assessment

Materials

- *Community Geography: GIS in Action* book (one per student)
- Computers for all students (for the exercise and assessment)

Student handouts from this lesson to be photocopied:

- Assessment
- Redlands map handout

Standards

GEOGRAPHY STANDARD	MIDDLE SCHOOL	HIGH SCHOOL
1 How to use maps and other geographic representations, tools, and technologies to acquire, process, and report information from a spatial perspective	The student knows and understands how to make and use maps, globes, graphs, charts, models, and databases to analyze spatial distributions and patterns.	The student knows and understands how to use maps and other graphic representations to depict geographic problems.
3 How to analyze the spatial organization of people, places, and environments on Earth's surface	The student knows and understands how to use spatial concepts to explain spatial structure.	The student knows and understands how to apply concepts and models of spatial organization to make decisions.
4 The physical and human characteristics of places	The student knows and understands how different human groups alter places in distinctive ways.	The student knows and understands the meaning and significance of place.

Objectives

The student is able to:

- Add data to a map
- Explore the attributes of the data
- Construct a map based upon predetermined criteria
- Create a map layout and print it

GIS skills and tools

▦ Add themes to the view		ⓘ Learn more about a selected record	
▦ Rename a theme		ᴬᵇ Change the font size of labels	
▦ Open an attribute table		▦ Label a feature	
▦ Clear selected features		▶ Select a label in the view and a graphic in the layout	
⊕ Zoom in on the view		✋ Pan the view to see different areas of the map	
⊖ Zoom out of the view		𝕋 Add text to a map layout	
◈ Zoom to the previous extent of the view			

GEOGRAPHIC INQUIRY STEP	GIS SKILL
Ask a geographic question	• Recognize and understand geographic questions posed in a scenario.
Acquire geographic resources	• Open project with street data and landmarks. • Add lodging and coffee-shop data to the project.
Explore geographic data	• Observe spatial relationship between themes. • Explore attributes using the Identify tool and attribute table. • Label streets, hotels, motels, and coffee and dessert shops.
Analyze geographic information	• Create a layout of downtown Redlands for a specific audience.
Act on geographic knowledge	• Print the layout for the Redlands visitors.

Teacher notes

Lesson introduction

Begin the lesson with a display and discussion of various map types such as topographic, street, and relief. Ask the students to identify the type of map that would be most useful to those needing to find their way around a new city. Next, have the students study a typical street map of their area and brainstorm what makes that map informative and usable. Record their ideas on the board. Ask students to focus on the specific components of maps that are most useful (labeling, symbols, north arrow, scale, title of the map, author of the map, and date the map was created).

In this GIS lesson, students will be led through the production of a street map using several layers of data provided for them. Distribute *Community Geography: GIS in Action* to each student or student group. Instruct the students to read the introductory materials for the module and the introductory exercise very carefully.

Exercise

Before completing this lesson with students, we recommend that you complete it as well. Doing so will allow you to modify the activity to accommodate the specific needs of your students.

TEACHER TIPS

✓ Have the students save their projects regularly. To do that, they should make the Project window active, click File, Save Project As, and then navigate to a drive where they have a folder to save their work.

✓ When renaming the project, advise students to use a short but descriptive name. For example, a student by the name of Maria Gomez could rename her project map_mg.apr, where "mg" represents her initials.

✓ Be sure students save their projects with unique names. If a student tries to save a new project with the same name as another project, the newest project will overwrite the old project.

✓ Project names cannot have spaces in them. ArcView will not recognize a project name longer than sixty characters or one with spaces.

Things to look for while the students are working on this activity:

• Are the students able to add new themes of data to their maps and identify the features on the map?

• Are the students able to change the map symbols?

• Are the students able to apply what they learned in the initial exercise to create an original map?

• Are the students answering the questions as they work through the procedure?

• Are the students creating maps that are clear and legible?

The students should be able to finish the exercise in one class period. If they don't, be sure to give them instructions on how to rename and save the project and record the new name so they can open it during the next class.

You can choose whether or not to have the students submit their final exercise map along with their assessment map for evaluation.

Conclusion

To conclude this lesson and introduce the assessment, engage the class in a discussion of the map-creation process. Be sure to have the students share the difficulties they had as well as the solutions they found. You can use this conversation as a springboard to explain the assessment.

Prior to distributing the Redlands map handout and assessment to the students, you must photocopy them and customize each handout for each student or student group. Note: The Redlands map handout sheet has two questionnaires per page and can be cut in half after copying to save paper. Each student or student group will receive a handout that has different data layers checked off and an assessment handout. Each handout will have a different combination of data that the students will have to map.

Assessment

This assessment will illustrate to you whether the students can apply the knowledge they learned in the exercise to create a map with a specific set of criteria. They will use the GIS skills they learned in the exercise to add data and label their map.

After you distribute the Redlands map handout and assessment to each student or student group, remind them to staple their handout to the map they submit for evaluation. Without the handout, you will not be able to evaluate whether they prepared the correct type of map. This assessment should be completed within one class period.

Middle school: Highlights skills appropriate to grades 5 through 8

The middle school assessment asks students to create and print a map. They must add themes, change theme symbols, zoom in, and label streets and places to satisfy a list of predetermined criteria. The students will write a brief paragraph describing how their thematic map will help the teacher group that requested the map.

High school: Highlights skills appropriate to grades 9 through 12

The high school assessment asks students to create and print a map. They must add themes, change theme symbols, zoom in, and label streets and places to satisfy a list of predetermined criteria. The students will write an essay on how their thematic map satisfies the needs of the teacher group that requested the map. It will include a discussion of other map types that could help the group leader plan a fun and informative trip.

REDLANDS MAP

Name _____ Date _____

Reminder: Staple this handout to your completed map.

A group of history teachers is visiting the Lincoln Memorial Shrine in Redlands, California. You are assigned to prepare a map of Redlands with the following information:

- ☐ Banks
- ☐ Coffee and dessert shops
- ☐ Landmarks
- ☐ Lodging

- ☐ Pizza parlors
- ☐ Restaurants
- ☐ Stores
- ☐ Other historical landmarks

- -

REDLANDS MAP

Name _____ Date _____

Reminder: Staple this handout to your completed map.

A group of history teachers is visiting the Lincoln Memorial Shrine in Redlands, California. You are assigned to prepare a map of Redlands with the following information:

- ☐ Banks
- ☐ Coffee and dessert shops
- ☐ Landmarks
- ☐ Lodging

- ☐ Pizza parlors
- ☐ Restaurants
- ☐ Stores
- ☐ Other historical landmarks

MIDDLE SCHOOL ASSESSMENT

Explore and label community features data for a city visitors map

Name _____ Date _____

Part 1

In this part of the assessment activity, you will create a map that is based on a handout your teacher gives you. It has been filled out by a leader of a group of history teachers who are visiting the Lincoln Memorial Shrine in Redlands, California. The handout will give you the information you need to accurately complete your map. Use your creativity and good planning to make an attractive and informative map. You will create a layout and print your map. The final map should include the following components:

- Map
- Text labels or descriptions
- Title
- Orientation (compass rose)
- Author (your name)
- Date

Part 2

In part 2 of the assessment, write a brief paragraph that describes how your map satisfies the criteria on the handout and how your map will be helpful to the leader.

Address the following questions when writing your paragraph:

- How did the symbols you chose give a clear representation of the features on the map?
- Did you use the zoom and pan tools to effectively choose the area to put on the map?
- When the leader looks at the map, do they have a good idea of how they might use this map to get their group around? Must they go by car or bus or can they walk?
- Did you put anything on the map that the leader did not ask for, but which might improve the total informational value of the map for the leader?

MIDDLE SCHOOL ASSESSMENT RUBRIC

Explore and label community features data for a city visitors map

STANDARD	EXEMPLARY	MASTERY	INTRODUCTORY	DOES NOT MEET REQUIREMENTS
1 The student knows and understands how to make and use maps, globes, graphs, charts, models, and databases to analyze spatial distributions and patterns.	Creates a map that accurately displays the requested information and other information useful to the end user. Includes all the components of a good map: appropriate labels and symbology, title, north arrow, author, and date.	Creates a map that accurately displays the requested information. Includes all the components of a good map, including appropriate labels and symbology, title, north arrow, author, and date.	Creates a map that displays some or all of the requested information. Includes some of the components of a good map, such as labels and symbology, title, north arrow, author, and date.	Map does not accurately display the requested information. Includes only a few key map components.
3 The student knows and understands how to use spatial concepts to explain spatial structure.	The layout is presented in an appropriate scale, and may also include variations on the scale to clarify key areas. There is significant spatial information with little or no other verbal or written explanation needed.	The printed map is presented in the appropriate scale and displays adequate spatial information with little or no other verbal or written explanation needed.	The printed map is presented in the appropriate scale, but does not have adequate spatial information. Some explanation is needed to provide adequate understanding of the map.	The printed map is not presented in an appropriate scale and does not have adequate spatial information. It is difficult to understand without further explanation from the student.
4 The student knows and understands how different human groups alter places in distinctive ways.	The written essay explains how the student-created map illustrates the human characteristics of the area in a variety of contexts. Analyzes the effectiveness of the map to the end user.	The written essay explains how the student-created map illustrates the human characteristics of the area in a variety of contexts.	The written essay explains how the student-created map illustrates the human characteristics of the area in one or two contexts.	Has difficulty explaining how the map illustrates the human characteristics of a place.

This is a four-point rubric based on the National Standards for Geographic Education. The "Mastery" level meets the target objective for grades 5–8.

HIGH SCHOOL ASSESSMENT

Explore and label community features data for a city visitors map

Name _____   Date _____

Part 1

In this part of the assessment activity, you will create a map that is based on a handout your teacher gives you. It has been filled out by a leader of a group of history teachers who are visiting the Lincoln Memorial Shrine in Redlands, California. The handout will give you the information you need to accurately complete your map. Use your creativity and good planning to make an attractive and informative map. You will create a layout and print your map. The final map should include the following components:

- Map
- Text labels or descriptions
- Title
- Orientation (compass rose)
- Author (your name)
- Date

Part 2

In part 2 of the assessment, write an essay that describes how your map satisfies the criteria on the handout and how your map will be helpful to the leader. Include information on other types of maps that could aid a group leader in creating a fun and informative trip for the group's members.

Some questions to consider when writing your essay are:

- How did the symbols you chose give a clear representation of the features on the map?
- Did you use the zoom and pan tools to effectively choose the area to put on the map?
- When the leader looks at the map, do they have a good idea of how they might use this map to get their group around? Must they go by car or bus or can they walk?
- Did you put anything on the map that the leader did not ask for, but which might improve the total informational value of the map for the leader?
- What other data would be useful in producing additional maps for the group leader?

HIGH SCHOOL ASSESSMENT RUBRIC

Explore and label community features data for a city visitors map

STANDARD	EXEMPLARY	MASTERY	INTRODUCTORY	DOES NOT MEET REQUIREMENTS
1 The student knows and understands how to use maps and other graphic representations to depict geographic problems.	Selects the best data to create the requested map, and provides the end user additional features to enhance the trip. Includes all the components of a good map, including appropriate labels and symbology, title, north arrow, author, and date.	Selects the best data to create the requested map. Includes all the components of a good map, including appropriate labels and symbology, title, north arrow, author, and date.	Selects the best data to create the requested map. Includes some of the components of a good map, such as appropriate labels and symbology, title, north arrow, author, and date.	Data selected provides little or none of the requested information. Includes some of the components of a good map, such as appropriate labels and symbology, title, north arrow, author, and date.
3 The student knows and understands how to apply concepts and models of spatial organization to make decisions.	The layout is presented in an appropriate scale, and includes variations on the scale to clarify key areas. There is adequate spatial information with little or no other verbal or written explanation needed. Allows the end user to easily plot out the trip.	The printed map is presented in the appropriate scale and displays adequate spatial information with little or no other verbal or written explanation needed. Allows the end user to easily plot out the trip.	The printed map is presented in the appropriate scale, but does not have adequate spatial information. The end user may have difficulty plotting the trip.	The printed map is not presented in an appropriate scale and does not have adequate spatial information. It is difficult to understand without further explanation from the student.
4 The student knows and understands the meaning and significance of place.	Explains in a written essay other possible data or maps that could be useful to the end user. Provides adequate evidence for the new maps in reference to the human characteristics of the place and creates sample maps to illustrate this point.	Explains in a written essay other possible data or maps that could be useful to the end user. Provides adequate evidence for the new maps in reference to the human characteristics of the place.	Explains in a written essay other possible data or maps that could be useful to the end user. Provides little or no evidence for the new maps in reference to the human characteristics of the place.	Explains in a written essay how the student-produced map is useful to the end user, but does not provide other data or map possibilities.

This is a four-point rubric based on the National Standards for Geographic Education. The "Mastery" level meets the target objective for grades 9–12.

Module 2: Reducing crime

Lesson

Case study: Deciding where to increase neighborhood police patrols
Exercise: Geocode crime data to map and analyze robbery hot spots

Lesson overview

Students will investigate crime data to determine where and when various types of robberies are occurring. They will learn the skill of geocoding addresses to map the crime data from a tab-delimited text file. Then they will identify criminal hot spots and determine which police beats are in need of increased patrols.

Estimated time

45-MINUTE CLASS	ACTIVITY
1	Exercise introduction
	Exercise part 1
2	Exercise part 2
3–5	Assessment

Materials

- *Community Geography: GIS in Action* book (one per student)
- Computers for all students (for the exercise and assessment)
- Color printer (to print the student maps)
- Transparency: ArcTown

Student handouts from this lesson to be photocopied:

- Assessment

Standards

GEOGRAPHY STANDARD	MIDDLE SCHOOL	HIGH SCHOOL
1 How to use maps and other geographic representations, tools, and technologies to acquire, process, and report information from a spatial perspective	The student knows and understands how to make and use maps, globes, graphs, charts, models, and databases to analyze spatial distributions and patterns.	The student knows and understands how to use geographic representations and tools to analyze, explain, and solve geographic problems.
3 How to analyze the spatial organization of people, places, and environments on Earth's surface	The student knows and understands how to use the elements of space to describe spatial patterns.	The student knows and understands the spatial behavior of people.
5 That people create regions to interpret Earth's complexity	The student knows and understands the elements and types of regions.	The student knows and understands how to use regions to analyze geographic issues.
18 How to apply geography to interpret the present and plan for the future	The student knows and understands how to apply the geographic point of view to solve social problems by making geographically informed decisions.	The student knows and understands how to use geographic knowledge, skills, and perspectives to analyze problems and make decisions.

Objectives

The student is able to:

- Explain what geocoding is and how it works
- Use GIS technology to create a map of crime locations from a table containing street addresses
- Map and analyze patterns within the crime data and suggest the best areas for increased police patrols

GIS skills and tools

⊕	Zoom in on the map	↗	Zoom to the active theme
❶	Identify a feature	T	Add text to the map
◈	Zoom the map to the previous extent	A&C	Access the Font Palette
▤	Open the Theme Properties dialog	☐	Select features
?	Locate an address	▤	Sort a table in ascending order
▶	Select a graphic	▦	Promote selected records
▦	Open a theme attribute table	Σ	Summarize an attribute field
○	Draw a circle	▣	Clear features that are selected

GEOGRAPHIC INQUIRY STEP	GIS SKILL
Ask a geographic question	• Recognize and understand geographic questions posed in a scenario.
Acquire geographic resources	• Add a table of crime data to the project.

Explore geographic data	• Zoom in to see detailed features.
	• Observe the spatial relationships between themes.
	• Explore attributes using the Identify tool and attribute table.
	• Geocode crime data.
	  – Make a theme matchable.
	  – Locate an address.
	  – Match a list of addresses.
	  – Interactively review unmatched addresses.

Analyze geographic information	• Visually analyze spatial patterns of crimes in relation to other crimes and geographic features. • Classify data by attribute. • Add a specialized marker set to the symbol palette. • Create a new shapefile and add features to it. • Visually identify a crime cluster and delineate it with a circle. • Select features in one theme (crimes) located within a feature in another theme (hot spot). • Calculate summary statistics for a set of selected features. • Compare and contrast crime hot spots by interpreting the summary statistics.
Act on geographic knowledge	• Recommend five beats for increased police patrols.

Teacher notes

Lesson introduction

Begin this exercise with a brief discussion about crime before the students begin working with ArcView on computers. How do we define crime? Where do crimes occur? When do they occur? Ask the students what they know about modern crime-fighting techniques from what they have seen on the news, television, and in movies. How do they think real law enforcement officials use technology?

Have the students read the case study "Deciding where to increase neighborhood police patrols" prior to completing the exercise. Discuss the case study and explain that they will use crime data from the Dallas Police Department very similar to that used by the students at Bishop Dunne Catholic School.

Part 1 of the exercise has the students geocode tabular crime data to a street shapefile. To help your students better understand the geocoding process before they work on the computer, lead them in the following nontechnical geocoding activity.

Human geocoder activity

The purpose of this activity is to introduce the concept of geocoding. It will bring your students' attention to the various elements of street addresses and the importance of complete and accurate data when address matching.

1 Define the word *geocode*. Geocode means to match addresses in a list (e.g., a table) to locations on a map (e.g., a shapefile) so that the items in the list can be displayed as points on the map. Explain to the students that they will play the role of human geocoders.

2 Display the ArcTown map transparency in this section and orient students to it. The street map includes street names, street types, address ranges, and directional information for each street segment (the reference theme), and a list of addresses below the map that need to be located (the "address table").

3 Ask a volunteer to direct you to the location of the first address on the map. Have the students explain why they decided on that location. You may need to coach the students through the initial process.

4 When the students explain the geocoding process they should take into account the individual parts of an address (e.g., number, street name, prefix, suffix).

5 Repeat this process with the other addresses in the list.

Note: The first two addresses are correct and on the map, but the other two have common problems, either with the street map or the address table (e.g., a missing address, out of range, no prefix).

To extend the human geocoder exercise:

• Create a transparency of your own community to use for the human geocoder activity using the ArcTown transparency as a model.

• After doing the human geocoder activity, have students generate a list of addresses from a printed or online telephone book and map them on a paper map of your own community. For example, they could map all grocery stores, libraries, or high schools.

Exercise

Before completing this lesson with students, we recommend that you complete it as well. Doing so will allow you to modify the activity to accommodate the specific needs of your students.

Read the exercise scenario aloud with the class. Explain that in this activity they will be asked to recommend five beats for the robbery task force patrols in the Dallas Police Department's southwest division. Before they can analyze the robbery locations, they will need to use the GIS to geocode addresses in a text file containing the robbery data. They will map the crime data according to the type of robbery (business, residential, or individual), identify hot spots for criminal activity, and then summarize the number and types of robberies in order to come up with their recommendations. Clarify any questions before the students begin to work individually.

TEACHER TIPS

✓ You should have the students save their projects regularly, especially at the end of part 1 and part 2 of the exercise. More information about saving and renaming projects can be found in the module 1 "Teacher notes."

✓ It is best to do part 1 in a single class period because it is important to complete the entire geocoding process in one ArcView session.

Things to look for while students are working on this activity:

- Are the students saving their work periodically to the designated directory?
- Are the students using a variety of tools when exploring the map (zoom tools and so on)?
- Are the students able to shift between tables and views, and resize windows appropriately?
- Are the students developing additional geographic questions based on their findings?

Conclusion

Once students have finished the exercise, have them save the project so they may refer to it for the assessment. Ask them to share their task force patrol list with the class. Lead a brief discussion comparing their recommendations. Did everyone come to the same conclusions? Student conclusions are likely to vary, in part because their hot spots will vary in location and size (there are more than five possible robbery clusters), and in part because their logic in choosing priority beats may vary. Explain that in the assessment they will be asked to create a map of their recommendations by watch period to present formally to the class.

Assessment

Assign the students to work in teams of three for the assessment. Students will create a map and a formal presentation to the class as though their classmates were representatives of the Dallas Police Department. The map will be created from the exercise data and analysis, so students will need access to the computers and, optionally, to a printer.

Students will analyze the robbery data by watch using the analysis techniques they learned in the exercise (drawing circles around hot spots, summarizing hot spots by number and type of robbery). They will use the shapefiles for each of the three watches that are included in the exercise data folder.

Ideally, the final presentations should be created in a multimedia presentation tool so they can be viewed easily by the entire class. The views can be exported from ArcView into a graphic file format compatible with a variety of presentation tools. See the "Act on geographic knowledge" section of "On your own: Project planning" in the *Community Geography: GIS in Action* book for additional strategies for presentations.

There are a number of alternative formats for submitting the final student assessments:

- Multimedia presentation
- Illustrated brochure
- Illustrated text (word processor) document

Middle school: Highlights skills appropriate to grades 5 through 8

Assign each student group to a particular watch period (a space to write the assigned watch is provided on the assessment sheet). Each group will create a hot-spot map similar to the one produced in the exercise to identify concentrated areas of robberies for their assigned watch. They will prepare and give a formal presentation to the class explaining the process for identifying the hot spots and why these particular locations need increased patrols by the Robbery Task Force.

High school: Highlights skills appropriate to grades 9 through 12

Each student group must analyze all three watch periods and create two hot spots for each watch. They will summarize the number and type of robberies in each of the six hot spots. Each group will create three layouts, one for each watch. Finally, each group will create a formal presentation to the class explaining the process for identifying the hot spots and why these particular locations need increased patrols by the Robbery Task Force.

HUMAN GEOCODER ACTIVITY

ArcTown, USA

		Lake St.		Prospect St.		Grand	
						Ave.	
200	198 **W. Main** St. 100		100 **E. Main St.** 198			200	298
201	199 101		101 199			201	299
			North Ave.				
200	198 **W. State St.** 100		100 **E. State St.** 198			200	
201	199 101		101 199			201	

150 W. State St.
251 E. Main St.
105 Grand Ave.
320 Main St.

 N

ANSWER KEY: HUMAN GEOCODER ACTIVITY

ArcTown, USA

● **150 W. State St.** *(Even side of street)*
■ **251 E. Main St.** *(Odd side of street)*
105 Grand Ave. *(Not matchable: no address ranges on map)*
320 Main St. *(Not matchable: out of range on map and no directional suffix in table)*

N

MIDDLE SCHOOL ASSESSMENT

Geocode crime data to map and analyze robbery hot spots

Name _____   Date _____

The Dallas Police Department was pleased with your analysis of all robberies in the southwest division. Now it wants a similar analysis of robberies broken down by watch times. You and your team have been asked to analyze robberies for your assigned watch time and then present your results to the Robbery Task Force.

Your assigned watch:

_____

Dallas Police Department patrol times

WATCH	TIME
Watch 1	12:00 midnight to 8:00 A.M.
Watch 2	8:00 A.M. to 4:00 P.M.
Watch 3	4:00 P.M. to 12:00 midnight

Part 1. Create a crime map

You will create a map that illustrates three robbery hot spots for your assigned watch using a new shapefile that has been created for you. Be prepared to explain in your presentation how you identified the hot spots in your map and what type of robbery occurs at each hot spot during your assigned watch.

A Using the ArcView project and data from the exercise, first create a new view in your project with the following themes:

THEME NAME	DATA SOURCE
SW Beat Zones	sw_beats.shp
Roads	sw_roads.shp
Arterials	sw_arterials.shp
All robberies	crime_abc.shp
Watch 1 Robberies	watch1.shp
Watch 2 Robberies	watch2.shp
Watch 3 Robberies	watch3.shp

B Following steps 3–14 in part 2 of the exercise, create a new shapefile containing three robbery hot spots for your assigned watch. Label your hot spots A, B, and C. If you wish to classify the robberies in your watch theme, remember that you can load the Offense.avl legend you saved in the exercise.

C Create a layout of your watch hot spots. Be sure to include the following information:
- Title
- Orientation (compass rose)
- Author(s)
- Date

Challenge

Create a layout comparing your watch hot-spot map to other watches.

Part 2. Present your findings

Prepare and give a presentation to the police department (your class) explaining why you think assigning additional patrols to your hot spots will be the most effective. Show your map, and be sure to address the following items in your presentation:

- Where do most of the crimes occur on your watch compared to where they occur on all three watches together?
- On which beats do you think the police should focus their patrols based on your hot spots?
- Which types of robbery are most prevalent on your watch and why do you think this is so?
- What results do you predict for the robbery task force if your plan is implemented?

MIDDLE SCHOOL ASSESSMENT RUBRIC

Geocode crime data to map and analyze robbery hot spots

STANDARD	EXEMPLARY	MASTERY	INTRODUCTORY	DOES NOT MEET REQUIREMENTS
1 The student knows and understands how to make and use maps, globes, graphs, charts, models, and databases to analyze spatial distributions and patterns.	The student creates a legible map that meets all the listed requirements (appropriate themes, title, north arrow, and so on) and clearly identifies three robbery hot spots on the assigned watch. The student may also create an additional map comparing their watch to others.	The student creates a legible map that meets all the listed requirements (appropriate themes, title, north arrow, and so on) and clearly identifies three robbery hot spots on the assigned watch.	The student creates a map that meets most of the listed requirements (appropriate themes, title, north arrow, and so on) and identifies two to three robbery hot spots on the assigned watch.	The student creates a map that meets some of the listed requirements and identifies one to two robbery hot spots on the assigned watch.
3 The student knows and understands how to use the elements of space to describe spatial patterns.	Through visual and quantitative analysis of the data, the student summarizes robbery types for the hot spots they have identified on the assigned watch.	Through visual analysis of the data, the student summarizes robbery types for the hot spots they have identified on the assigned watch.	Through visual analysis of the data, the student attempts to summarize robbery types on the assigned watch.	The student has difficulty identifying different robbery types on the assigned watch.
5 The student knows and understands the elements and types of regions.	The student identifies crime hot spots, recommends beats on the assigned watch that need increased patrols, and notes which robbery type is most prevalent.	The student identifies crime hot spots and recommends beats on the assigned watch that need increased patrols.	The student recommends areas (other than beats) that need increased patrols on the assigned watch.	The student has difficulty determining areas that need increased patrols on the assigned watch.
18 The student knows and understands how to apply the geographic point of view to solve social problems by making geographically informed decisions.	The student presents a clear argument as to why their solution is the most logical based on the data and makes predictions on the impact of the focused patrols on crime.	The student presents a good argument as to why their solution is the most logical based on the data.	The student attempts to justify their solution, but does not provide ample evidence from the data.	The student has difficulty explaining their thinking and provides little or no evidence for the solution.

This is a four-point rubric based on the National Standards for Geographic Education. The "Mastery" level meets the target objective for grades 5–8.

HIGH SCHOOL ASSESSMENT

Geocode crime data to map and analyze robbery hot spots

Name _____   Date _____

The Dallas Police Department was pleased with your analysis of all robberies in the southwest division. Now you and your team have been selected by the department to do a more detailed analysis of the robberies by watch times and then report your results.

Dallas Police Department patrol times

WATCH	TIME
Watch 1	12:00 midnight to 8:00 A.M.
Watch 2	8:00 A.M. to 4:00 P.M.
Watch 3	4:00 P.M. to 12:00 midnight

Part 1. Create a crime map series by watch

You will create a map that illustrates three robbery hot spots for each watch using a new shapefile that has been created for you. Be prepared to explain in your presentation how you identified the hot spots in your map and what type of robbery occurs at each hot spot during each watch.

A Using the ArcView project and data from the exercise, first create a new view in your project with the following themes:

THEME NAME	DATA SOURCE
SW Beat Zones	sw_beats.shp
Roads	sw_roads.shp
Arterials	sw_arterials.shp
All robberies	crime_abc.shp
Watch 1 Robberies	watch1.shp
Watch 2 Robberies	watch2.shp
Watch 3 Robberies	watch3.shp

B Following the steps in part 2 of the exercise, create a new hot-spot shapefile for watch 1 and draw two hot spots around watch 1 robbery clusters. Label your hot spots 1A and 1B. (Note: you can save time when classifying the robberies in the watch themes by loading the Offense.avl legend you saved earlier.) Repeat the procedure, creating shapefiles and hot spots for watches 2 and 3. As you work, enter your analysis results in the "Summary of hot spots by watch" table.

C Create three layouts, one for each watch, illustrating your hot spots. Be sure to include the following information:
- Text labels or descriptions
- Titles
- Orientation (compass rose)
- Author(s)
- Date

Challenge

Create charts or graphs that illustrate the number and types of crime for each watch.

Part 2. Present your findings

Prepare and give a presentation to the police department explaining why you think assigning additional patrols to your hot spots will be the most effective. Show your maps, and be sure to address the following items in your presentation:

- Explain how you identified the hot spots in your map.
- Which types of robbery are most prevalent on each watch and why do you think this is so?
- Are the hot spots found in the same location across the different watch periods, or do the hot spots move from watch to watch?
- On which beats do you think the police should focus their patrols based on your analysis of the data across various watches?
- What results do you predict for the robbery task force if your plan is implemented?

Summary of hot spots by watch

	SHAPEFILE NAME	HOT SPOT	IDs OF BEATS IN HOT SPOT	BUSINESS ROBBERIES	INDIVIDUAL ROBBERIES	RESIDENTIAL ROBBERIES	TOTAL ROBBERIES
Watch 1	_____ .shp	1A					
		1B					
Watch 2	_____ .shp	2A					
		2B					
Watch 3	_____ .shp	3A					
		3B					

HIGH SCHOOL ASSESSMENT RUBRIC

Geocode crime data to map and analyze robbery hot spots

STANDARD	EXEMPLARY	MASTERY	INTRODUCTORY	DOES NOT MEET REQUIREMENTS
1 The student knows and understands how to use geographic representations and tools to analyze, explain, and solve geographic problems.	The student creates a map series (one map for each watch) that meets all the listed requirements (appropriate themes, title, orientation, and so on) and clearly identifies two robbery hot spots on each watch. They create graphs or charts further illustrating the number and types of crime by watch.	The student creates a map series (one map for each watch) that meets all the listed requirements (appropriate themes, title, orientation, and so on) and clearly identifies two robbery hot spots on each watch.	The student creates a map series (one map for each watch) that meets some of the listed requirements and attempts to identify two to three robbery hot spots on any watch.	The student creates one map rather than a series of maps illustrating hot spots without analyzing each individual watch.
3 The student knows and understands the spatial behavior of people.	The student hypothesizes why certain times of day have higher crime rates (or a higher percentage of a certain crime) than others and provides ample evidence (such as maps or charts) for their conclusions.	The student hypothesizes why certain times of day have higher crime rates (or a higher percentage of a certain crime) than others and provides some evidence for their conclusions.	The student attempts to hypothesize why certain times of day have different crime rates than others, but does not provide evidence for their thinking.	The student has difficulty identifying differences in crime and crime rates on each watch.
5 The student knows and understands how to use regions to analyze geographic issues.	In their maps, the student identifies various beats and watches they believe need increased patrols based on their analysis of the spatial data. Presents which crimes are most prevalent in the identified regions.	The student identifies various beats and watches they believe need increased patrols based on their analysis of the spatial data. Presents which crimes are most prevalent in the identified regions.	The student identifies beats or watches they believe need increased patrols based on their analysis of the spatial data.	The student attempts to identify beats or watches that need increased patrols, but does not provide evidence of their analysis.
18 The student knows and understands how to use geographic knowledge, skills, and perspectives to analyze problems and make decisions.	The student presents a clear argument as to why their solution is the most logical based on the data and makes predictions about the impact of the focused patrols on crime. The predictions are illustrated in additional maps created by the student.	The student presents a clear argument as to why their solution is the most logical based on the data and makes predictions about the impact of the focused patrols on crime.	The student presents a good argument as to why their solution is the most logical based on the data.	The student attempts to justify their solution, but does not provide ample evidence from the data.

This is a four-point rubric based on the National Standards for Geographic Education. The "Mastery" level meets the target objective for grades 9–12.

On your own

Overview

This section provides guidelines and information to help you implement a similar project in your own classroom. Crime analysis allows students to take a proactive stance on crime in their own community. It provides them with the opportunity to become part of a solution that will make them feel safer at school and at home. Young people sometimes feel as though they are unfairly portrayed as being part of problems such as graffiti or drug abuse within schools. Conducting a crime project on an issue that is relevant to themselves and their school will empower them to make a difference.

Estimated time

The length of time needed for a GIS crime-mapping project can vary, depending on how large a problem you intend to tackle. In the Bishop Dunne case study, the class worked on the project for the first semester of school.

Teacher tips on setting a time frame

- Start with a smaller, focused project that can be done in one to two weeks of class time.
- Use the small project to gain experience with the analysis software and methods you want to use for a larger project. For example, if you plan to use the ArcView Crime Analysis Application or another extension, use a smaller project to learn how it works and what you can do with it.
- Once you have gained some experience analyzing crime data, you and your class may want to establish long-term partnerships with your local law enforcement agency or community crime watch group.

Materials

- Computers and appropriate GIS software for all students
- Access to a color printer or plotter to print out maps (optional)
- Crime data
- Geocodable, current street data
- Class e-mail address for communication with community partners
- Reserved disk space for data storage

Standards

GEOGRAPHY STANDARD	MIDDLE SCHOOL	HIGH SCHOOL
1 How to use maps and other geographic representations, tools, and technologies to acquire, process, and report information from a spatial perspective	The student knows and understands how to make and use maps, globes, graphs, charts, models, and databases to analyze spatial distributions and patterns.	The student knows and understands how to use geographic representations and tools to analyze, explain, and solve geographic problems.
2 How to use mental maps to organize information about people, places, and environments in a spatial context	The student knows and understands how to translate mental maps into appropriate graphics to display geographic information and answer geographic questions.	The student knows and understands how mental maps reflect the human perception of places.
3 How to analyze the spatial organization of people, places, and environments on Earth's surface	The student knows and understands how to use the elements of space to describe spatial patterns.	The student knows and understands the spatial behavior of people.
5 That people create regions to interpret Earth's complexity	The student knows and understands the elements and types of regions.	The student knows and understands how to use regions to analyze geographic issues.
6 How culture and experience influence people's perception of places and regions	The student knows and understands how personal characteristics affect people's perception of places and regions.	The student knows and understands why different groups of people within a society view places and regions differently.
18 How to apply geography to interpret the present and plan for the future	The student knows and understands how to apply the geographic point of view to solve social problems by making geographically informed decisions.	The student knows and understands how to use geographic knowledge, skills, and perspectives to analyze problems and make decisions.

Objectives

The student is able to:

- Create a map of data from a table of statistics
- Recognize spatial patterns within the crime data
- Create maps outlining areas of high crime or needing attention
- Generate summary statistics and record their findings
- Recommend actions that will help reduce crime in the community

GIS skills and tools

GEOGRAPHIC INQUIRY STEP	GIS SKILL
Ask a geographic question	• Develop one or more geographic questions related to crime or crime prevention in your community.
Acquire geographic resources	• List data needed to answer the geographic questions. • Identify and obtain data from reliable sources. • Add crime data to the GIS project.

Explore geographic data	• Observe the spatial relationships within and between themes.
	• Explore attributes using the Identify tool and attribute table.
	• Geocode crime data.
	    – Make a theme matchable.
	    – Locate an address.
	    – Match a list of addresses.
	    – Correct any unmatched addresses.

Analyze geographic information	• Classify crime data by attribute (e.g., by offense, time of day, severity of crime). • Create a shapefile containing new polygons drawn around clusters of crime points. • Create buffers or measure distances from features such as parks or schools. • Generate and record summary statistics.
Act on geographic knowledge	• Create layouts and presentations that present findings and support the recommended actions.

Teacher notes

Ask a geographic question

Crime is a big subject and it's easy for one question or investigation to lead to another. You will want to make sure to choose one or more questions that you can investigate within the available time frame.

TEACHER TIPS

✓ Tell your students the time frame you have for this project.

✓ If you have a small window of time such as a couple of weeks, be sure to help your students narrow their focus.

✓ Make sure students think of all types of data and information they may need to analyze the crime issue they have chosen. The more specific and detailed they are in the planning stages, the easier the rest of the process will be for them.

✓ Have students brainstorm lists of important contact people who can help them in the process.

Acquire geographic resources

TEACHER TIPS ON DEVELOPING PARTNERSHIPS WITH LOCAL LAW ENFORCEMENT

✓ Once the students have developed their geographic question about crime, you may want to make a few phone calls on their behalf to develop partnerships. One of the first places to call is your community police department. Find out if they have a GIS department and explain that you are working with students who want to establish a partnership with them to study or fight crime in your community.

✓ Many police departments have neighborhood police liaisons who meet with community groups to solve such issues. Find out if your town has such a liaison and see if they would be willing to make a presentation and meet with your class.

TEACHER TIPS ON ACQUIRING CRIME DATA

✓ It is very important to communicate that students will be working with the data. Often police departments will clean the data to be sure that all privacy and related issues are eliminated from the data set they provide to you.

✓ You may want to request that they only provide you certain types of crime data (e.g., only robberies, or only vandalism). Be aware that this is real data and stress this point to the students.

✓ If your police department does not have any community partnership programs available, remember that you may obtain your data through Open Records as part of the Freedom of Information Act. More information can be found in *Community Geography: GIS in Action,* "On your own: Project planning."

Explore geographic data

Technical issues

- All mapped data in a GIS is in a map projection. A map projection is needed to fit the spherical earth onto a two-dimensional map on the computer screen. You should know the map projection of your spatial data. Data sets acquired from various sources may not always be in the same projection. ArcView has tools to reproject data so that it all lines up in your selected projection. See *Community Geography: GIS in Action,* "On your own: Project planning" for more information about handling data-projection issues.

- Have students practice the geocoding process by doing the human geocoder activity.

- It is helpful to have a good street map of the area or a contact at your local city planner's office to assist with questions on street addresses.

- Use your text resources as well. For example, the book *Getting to Know ArcView GIS* (ESRI Press, 2000) has detailed information and helpful definitions about geocoding. It also may help to have community GIS experts assist in geocoding the data.

- Decide how accurate the address matching needs to be. When geocoding, ArcView allows you to determine the level of accuracy by changing the spelling sensitivity or the minimum score required for an address to be considered a match.

Analyze geographic information

If you have a short time frame for your project, having students use a simple visual analysis technique is effective. They can complete a basic analysis of their data by following the process outlined in the exercise. This visual analysis technique has the students hypothesize about the location of hot spots by looking for clusters of dots.

Longer-term projects or projects with older students can take advantage of additional ArcView tools or extensions, such as the geoprocessing wizard found in ArcView 3.1 or higher, and the ArcView Spatial Analyst and ArcView Crime Analysis Application extensions. These tools will allow students to create density rasters and examine the data using quantitative techniques. When using these tools, multiple options are often available, such as specifying the raster cell size or the interpolation method to use when converting point data to a raster. Have your students explore the options to find those that result in the most useful and visually meaningful map.

The ArcView Spatial Analyst extension is available at a discounted price for an instructional site license from ESRI. The ArcView Crime Analysis Application extension, which requires ArcView Spatial Analyst, adds a series of wizards that make it easier to apply many of the ArcView Spatial Analyst functions to crime data. The Crime Analysis Application extension is free and can be downloaded from ESRI's Web site *(www.esri.com)* through the Law Enforcement Industry Solutions page.

The decision as to which type of analysis is best for your project is up to you and your students. Take a close look at your geographic question and determine which methodology will provide you with the best possible solution.

Act on geographic knowledge

The students can take a variety of actions based on their findings. One recommended action is to partner with law enforcement or other community crime-watch groups to implement a plan. This allows the students to actively participate with working professionals and teaches them the value of community partnerships. Stress to the students that they should not take this as an opportunity to attempt to track down thieves on their own. That is a job best left to the professionals. Refer to the "Act on geographic knowledge" section of "On your own: Project planning" in *Community Geography: GIS in Action* for ideas that can be used as springboards for taking safe action.

Acknowledgments

Thanks to ESRI instructor Jamie Parrish for contributing the human geocoder activity.

Lesson

Case study: Mapping noxious weeds
Exercise: Use GIS to map a leafy spurge infestation and compute its area

Overview

In this exercise, students will map GPS points representing a leafy spurge infestation in a farmer's field. Their task is to digitize circles of specific sizes around the GPS points and investigate the size of the infestation by using the summarize and calculate functions of ArcView. They will compute the area infested with leafy spurge and the area of the crop field, and calculate the percentage of the field that is infested.

Estimated time

45-MINUTE CLASS	ACTIVITY
1	Exercise introduction
2	Exercise part 1
3	Exercise part 2
4	Conclude exercise Begin assessment
5	Assessment

Materials

- *Community Geography: GIS in Action* book (one per student)
- Computers for all students (for the exercise and assessment)
- Color printer (to print the student maps)

Student handouts from this lesson to be photocopied:

- Assessment

Standards

GEOGRAPHY STANDARD	MIDDLE SCHOOL	HIGH SCHOOL
4 The physical and human characteristics of places	The student knows and understands how different physical processes shape places and how to analyze the physical characteristics of places.	The student knows and understands the changing physical and human characteristics of places.
8 The characteristics and spatial distribution of ecosystems on Earth's surface	The student knows and understands how human activities influence changes in ecosystems.	The student knows and understands the importance of ecosystems in people's understanding of environmental issues.
14 How human actions modify the physical environment	The student knows and understands the consequences of human modification of the physical environment.	The student knows and understands how to apply appropriate models and information to understand environmental problems.
18 How to apply geography to interpret the present and plan for the future	The student knows and understands how to apply the geographic point of view to solve social and environmental problems by making geographically informed decisions.	The student knows and understands how to use geographic knowledge, skills, and perspectives to analyze problems and make decisions.

Objectives

The student is able to:

- Add tabular GPS data with location coordinates and create an event theme of points
- Identify patterns of weed infestation relative to the physical characteristics of the earth noted on an aerial photograph
- Use math skills to convert acreages to circle radii to depict weed infestations
- Summarize and calculate the weed infestations in a farmer's field
- Compute the percentage of infestation compared to the farmer's field

GIS skills and tools

⊕	Zoom in on the map	○	Draw a circle
ⓘ	Identify features on a map	▶	Select a graphic
⛶	Zoom out from the map	▦	Open a theme's attribute table
▦	Change a theme name	▣	Calculate attribute values
⊕	Add a theme to the view	Σ	Summarize an attribute field
▨	Zoom to the active theme		

GEOGRAPHIC INQUIRY STEP	GIS SKILL
Ask a geographic question	• Recognize and understand geographic questions posed in a scenario.
Acquire geographic resources	• Enhance a basemap by adding landowner and road themes to a county map. • Add tabular GPS data to the project. • Create a new point shapefile and event theme containing the GPS location coordinates.

Explore geographic data	• Observe the spatial relationship of the field and invasive weeds by:
	  – Turning themes on and off
	  – Copying and pasting themes
	  – Changing theme names
	  – Rearranging the table of contents
	  – Digitizing circles around points
	  – Merging features
	  – Calculating the area of geographic features
	• Explore attributes using the Identify tool and attribute table.

Analyze geographic information	• Classify weeds by coverage and size of infestation. • Symbolize themes with unique values and graduated colors. • Spatially analyze the data to calculate percentage of the field's area infested by invasive weeds.
Act on geographic knowledge	• Produce a report for the farmer with the analysis findings so he can develop an eradication plan.

Teacher notes

Lesson introduction

This exercise encourages students to think about noxious weeds differently than they ordinarily might. Because noxious weeds are not "bad" if they are growing in their original habitat, it is important for students to recognize that these plants can be desirable in one ecosystem, yet be unwelcome in another.

Collect pictures of noxious weeds to share with your students. Check out the following sources:

- Internet—Do a search on noxious weeds, invasive weeds, or nonnative weeds.
- Library—Do a library search on the same keywords.
- Local extension agency or nature center, Bureau of Land Management office, or state department of natural resources—They will have many educational materials available for your use.

Hang up the pictures of noxious weeds across the board or around the room where students can look at them. Begin the lesson by spending a moment or two having the students comment on the pictures. It is possible that the only observation they make is that the flowers are pretty and look like wildflowers. After the students have commented briefly on the pictures, inform them that these are pictures of weeds.

Introduce the concept of a weed. What is a weed? Generally, people define a weed as a plant that is growing where it is not wanted. Point out to students that this definition reflects an entirely human viewpoint; we perceive the environment based on our values and goals.

Introduce the case study, "Mapping noxious weeds," by telling them it is about students who tackled the problem of noxious weeds in their community. Have the students read the case study in class.

Lead the class in a brief discussion with the purpose of drawing out information that the students picked up in the case study. Possible discussion questions include:

- How do noxious weeds appear in new locations? What role do human activities play in the relocation or spread of weed species?
- What did the students at Shelley High School do to combat weeds in their area?
- What are some of the adverse effects of noxious weeds on an area, including human, environmental, and economic effects?
- Where are noxious weeds found?
- Are there ways to rid an area of noxious weeds?
- Tell the students that throughout this lesson they will see several different terms used to refer to these weeds. The terms "noxious weeds," "invasive weeds," and "nonnative weeds" have subtle distinctions, but generally they refer to the same class of weeds.

Exercise

Before completing this lesson with students, we recommend that you complete it as well. Doing so will allow you to modify the activity to accommodate the specific needs of your students.

Read the exercise scenario aloud with the class. Explain that in part 1 they will use GIS to map GPS points in a table to an ArcView project and they will analyze noxious weed infestations across a county in Idaho. In part 2 they will digitize circles representing areas of noxious weed infestation in a particular farmer's field. They will use specific ArcView tools to summarize and calculate the data in order to compute the percentage of infestation of the farmer's field.

TEACHER TIPS

✓ Parts 1 and 2 are of unequal length (part 1 may take less time), but the students can progress from part 1 to part 2 if they have sufficient time before the end of class. If they begin part 2 and need to conclude a session when they are in the middle of editing a shapefile, they should stop editing and save their edits—they can complete the edits in the next session. (Have students note which step they left off at and whether or not they fully completed it.) Then they should save their ArcView project before ending the session.

✓ Tell students where they should save their ArcView projects and the shapefiles they will be creating. The exercise explicitly instructs students to save the project at part 1, step 13, and again at the end of parts 1 and 2. More information about saving and renaming projects can be found in the module 1 teacher notes.

✓ When the students digitize the circles around the infestation points in part 2, they will actually need to digitize only three circles, one for each category noted in part 1, step 28 of the exercise. Once they have digitized one size circle, they should copy and paste that circle to each of the other points in that category. That way, all circles representing the same infestation size will have exactly the same area. The point on the map in the fourth category ("other") should not be digitized because the infestation size is not known.

✓ Tell your students that they will not be able to match the size of the circles they digitize exactly to the size indicated in the table. They should do their best to get as close as possible, but could be as much as 1 or 2 feet off. They will be able to measure more precisely if they increase the map scale by zooming in or enlarging the view window.

✓ Advise students to read the instructions for each step very carefully, especially in part 2 where they summarize and calculate data. All steps must be followed carefully or they will have difficulty with the calculations.

✓ Remember that student results for the total area of weed infestation will vary because the size of the circles the students digitize will vary slightly, and the actual radius of the digitized circle will be used in all further calculations.

Things to look for while the students are working on this activity:

- Are the students changing the order of their table of contents to successfully display their maps?
- Are the students able to digitize the circles close to the approximate size and then correctly copy and paste the circles, or are they digitizing each individual circle?
- Are the students able to successfully calculate the area of the farmer's field?
- Are the students saving their work periodically to the designated folder?

Technical tip

- In the exercise (and assessment), students create circular polygons representing infestation areas. The radius lengths to use are given in a table. The following example shows how these radii were derived, in case you want to explain the process to your students or apply this technique to a different situation.

  1 Begin with the estimated infestation size (e.g., 3 acres).
  2 Convert the infestation size to square feet (because you will measure the radius in feet). One acre = 43,560 square feet; thus, $3 \times 43{,}560 = 130{,}680$ square feet.
  3 Use the formula for the area of a circle to determine the radius.

$$\text{area} = \pi r^2 \text{ (where } \pi = 3.14 \text{ and } r = \text{the radius)}$$
$$\sqrt{\text{area}/\pi} = r$$

Thus,

$$130{,}680/3.14 = 41{,}617.8$$
$$\sqrt{41{,}617.8} = 204 \text{ feet} = r$$

Conclusion

When the students have completed their project, make sure they save it so they can use it in the assessment. Conclude this lesson by having a class discussion centered around the following questions:

- What percentages of field infestation did the different students or student groups come up with? Record these numbers on the board.
- Were the percentages between students or groups consistent with each other or was there a large difference in percentages? (For example, one group calculates a 60-percent infestation of the farmer's field while another group calculates a 20-percent infestation.)
- If there were inconsistencies in the students' results, ask the students to come up with explanations as to why that might have occurred (e.g., math errors, radius size differences).
- How quickly do students think the weeds might spread if left unchecked?

Assessment

For both the middle school and high school assessments, students will be given a scenario where they will need to compute the percentage of leafy spurge infestation in a farmer's field after four years of unchecked growth. Students will be given a simple model to compute the growth of each infestation after four years, and will use the procedures they learned in the exercise to compute the change in area affected.

Refer to the module 3 answer key for selected assessment answers.

Middle school: Highlights skills appropriate to grades 5 through 8

Middle school students will compute the four-year projected infestation, map their results, and identify long-range consequences of unchecked leafy spurge growth. In their report to the farmer, students will recommend a method of leafy spurge abatement and explain why their chosen method is the best.

High school: Highlights skills appropriate to grades 9 through 12

High school students will compute the four-year projected infestation and map their results. After conducting research on the topic, students will prepare a report to the farmer that includes an explanation of different means of seed dispersal and how one of these means can be incorporated into the model. In addition, they will identify long-term consequences of leafy spurge infestation to the nonagricultural environments in eastern Idaho, and will recommend an eradication method for the farmer to use on his field. Students will be asked to provide a list of all bibliographic resources they used to prepare the report for the farmer.

MIDDLE SCHOOL ASSESSMENT

Use GIS to map a leafy spurge infestation and compute its area

Name _____ Date _____

The farmer would like you to help him predict the spread of leafy spurge in his field and to recommend the method he should use to slow the infestation. In this assessment, you will produce a map layout of his field and a written report that answers his questions about controlling the spread of weeds.

Part 1. Create and print a map layout

You will create and print a map layout of the farmer's field that shows the current extent of the leafy spurge infestation and the projected infestation after four years of uncontrolled growth.

1 With the information you have given the farmer about leafy spurge and its ability to spread seeds up to 15 feet away from the original plant, the farmer has developed a simple model for you to use. Calculate the radius of the infestation circles below by multiplying the number of years (four years) by the projected growth per year (15 feet). Then add the projected growth to the initial year radius to determine the projected radius after four years. Complete the table below.

	RADIUS OF INFESTATION IN FEET		
AVERAGE ACRES WITH LS	Initial year (feet)	Projected growth after four years	Projected radius size after four years
0.05	26		
0.55	87		
3	204		

2 Open your saved ArcView project from the exercise (weeds_abc.apr, where "abc" represents your initials). In part 2 of the exercise, use steps 1 through 11 as a guide to create a new theme of circles that represents the projected infestation area after four years. Draw circles using the radius sizes you calculated in the table above. (Do not draw a circle around the blue dot.)

3 In part 2 of the exercise, use steps 12 through 21 as a guide to merge the circles and calculate the total area of the leafy spurge infestation after four years.

   a Predicted total area of infestation in four years:

      _____ square meters or _____ acres

   b Percentage of the total field area taken up by the infestation in four years: _____ %

4 Create and print a map layout that shows both the current and four-year projection of leafy spurge infestation. Your layout should include all of the following elements:

- Current extent of leafy spurge infestation
- Four-year projection of leafy spurge infestation
- Leafy spurge GPS point locations
- Basemap information (aerial photograph and/or roads)
- Legend, scale, north arrow, title, author, and date

Part 2. Written report and recommendations

The farmer has prepared the following questions that he would like you to answer in your report. Write the answers on a separate sheet of paper.

1 If I am not able to stop the spread of leafy spurge in the field within the next couple of years, what long-range consequences do you foresee? Be sure to include a list of different types of consequences.

2 The following table lists four common methods used to control the spread of leafy spurge and some advantages and disadvantages of each method. Which one method do you recommend I use to control the leafy spurge on my property, and why? What are the advantages of your chosen method over the other methods listed?

METHOD	ADVANTAGES	DISADVANTAGES
Spray herbicide	• Works quickly • Repeated application may drastically reduce leafy spurge	• Expensive • May also harm crop plants or native plants • Must be applied at just the right time in the leafy spurge growth cycle
Beetles	• Free or nominal cost • Will not harm crop plants • Does not adversely affect the ecosystem	• Must be protected from insecticides, soil cultivation, burning, grazing, and so on for several years while they become established • Time intensive to collect • Can take five to ten years to become established and to have an effect • Will maintain leafy spurge at manageable levels but not eradicate
Sheep/Goats	• Leafy spurge is nontoxic and nutritious for sheep and goats • They prefer to eat flowering plants (like leafy spurge) and will leave the grasses to grow	• Must rotate animals among fields; it is costly to move them great distances • Cost of fencing • Need many animals to combat larger infestations • Seeds can get caught in hooves or hair and be transported to previously noninfested areas • Will maintain leafy spurge at manageable levels but not eradicate
Manual removal	• If done regularly for two to three years, may eradicate • No harm to other plants and grasses	• Plants must be pulled within seven days of germinating • If any of the root is left behind, it will grow again • Labor intensive • Only feasible in small, nonestablished infestations

MIDDLE SCHOOL ASSESSMENT RUBRIC

Use GIS to map a leafy spurge infestation and compute its area

STANDARD	EXEMPLARY	MASTERY	INTRODUCTORY	DOES NOT MEET REQUIREMENTS
4 The student knows and understands how different physical processes shape places and how to analyze the physical characteristics of places.	Creates and prints a map layout that includes all of the required map elements (e.g., north arrow, legend, scale). The map accurately and clearly displays current and projected leafy spurge infestation.	Creates and prints a map layout that includes most of the required map elements (e.g., north arrow, legend, scale). The map accurately displays current and projected leafy spurge infestation.	Creates and prints a map layout that includes some of the required map elements (e.g., north arrow, legend, scale). The map attempts to display current and projected leafy spurge infestation, but is inaccurate in places.	Creates and prints a map layout that includes few of the required map elements (e.g., north arrow, legend, scale). The map displays either the current or projected leafy spurge infestation, but not both.
8 The student knows and understands how human activities influence changes in ecosystems.	Recommends a weed abatement method and accurately describes its effect on the ecosystem of the field. The written explanation includes why this method is the most appropriate for the farmer and includes a detailed description of its advantages over other methods.	Recommends a weed abatement method and accurately describes its effect on the ecosystem of the field. The written explanation includes why this method is the most appropriate for the farmer.	Recommends a weed abatement method, but has difficulty describing its effect on the ecosystem of the field. The written explanation includes little information on why this method is the most appropriate for the farmer.	Recommends a weed abatement method, but provides little or no information on its effect on the ecosystem. Does not explain why this method is the most appropriate for the farmer.
14 The student knows and understands the consequences of human modification of the physical environment.	Provides a detailed description of various long-range consequences of unchecked leafy spurge infestation. Includes a comprehensive list of consequences and connects them to the farmer's field.	Provides a description of various long-range consequences of unchecked leafy spurge infestation. Includes a list of consequences and connects them to the farmer's field.	Provides some description of long-range consequences of unchecked leafy spurge infestation. Includes a list of consequences, but does not connect them to the farmer's field.	Includes very little description of long-range consequences of unchecked leafy spurge infestation. Does not connect the consequences to the farmer's field.
18 The student knows and understands how to apply the geographic point of view to solve social and environmental problems by making geographically informed decisions.	Creates accurate infestation projections and maps them. Uses the infestation projections to recommend an appropriate weed abatement method for the farmer. Explains how the recommended method will solve the farmer's leafy spurge infestation problem.	Creates accurate infestation projections and maps them. Uses the infestation projections to recommend an appropriate weed abatement method for the farmer.	Attempts to create infestation projections and maps them. Recommends a weed abatement method, but does not connect it to the farmer's situation.	Attempts to create infestation projections, but has difficulty mapping them. Recommends a weed abatement method, but does not connect it to the farmer's situation.

This is a four-point rubric based on the National Standards for Geographic Education. The "Mastery" level meets the target objective for grades 5–8.

HIGH SCHOOL ASSESSMENT

Use GIS to map a leafy spurge infestation and compute its area

Name _____ Date _____

The farmer would like you to help him predict the spread of leafy spurge in his field and to recommend the method he should use to slow the infestation. In this assessment, you will produce a map layout of his field and a written report that answers his questions about controlling the spread of weeds.

Part 1. Create and print a map layout

You will create and print a map layout of the farmer's field that shows the current extent of the leafy spurge infestation and the projected infestation after four years of uncontrolled growth.

1 With the information you have given the farmer about leafy spurge and its ability to spread seeds up to 15 feet away from the original plant, the farmer has developed a simple model for you to use. Calculate the radius of the infestation circles below by multiplying the number of years (four years) by the projected growth per year (15 feet). Then add the projected growth to the initial year radius to determine the projected radius after four years. Complete the table below.

	RADIUS OF INFESTATION IN FEET		
AVERAGE ACRES WITH LS	Initial year (feet)	Projected growth after four years	Projected radius size after four years
0.05	26		
0.55	87		
3	204		

2 Open your saved ArcView project from the exercise (weeds_abc.apr, where "abc" represents your initials). In part 2 of the exercise, use steps 1 through 11 as a guide to create a new theme of circles that represents the projected infestation area after four years. Draw circles using the radius sizes you calculated in the table above. (Do not draw a circle around the blue dot.)

3 In part 2 of the exercise, use steps 12 through 21 as a guide to merge the circles and calculate the total area of the leafy spurge infestation after four years.

   a Predicted total area of infestation in four years:

   _____ square meters or _____ acres.

   b Percentage of the total field area taken up by the infestation in four years: _____ %

4 Create and print a map layout that shows both the current and four-year projection of leafy spurge infestation. Your layout should include all of the following elements:

   • Current extent of leafy spurge infestation
   • Four-year projection of leafy spurge infestation
   • Leafy spurge GPS point locations
   • Basemap information (aerial photograph and/or roads)
   • Legend, scale, north arrow, title, author, and date

Part 2. Written report and recommendations

The farmer has prepared the following questions that he would like you to answer in your report. Answer the questions using information from the case study as well as from your own research using resources such as books and the Internet. Write your answers on a separate sheet of paper.

1 You have used a simple model to predict the spread of leafy spurge in my field. The model accounts for only one way that leafy spurge spreads—the natural dispersal of seeds by propulsion from existing plants. What other ways could these weeds be spread within my crop field, to other areas of my property, or even to areas beyond my property? Describe how you would change the model to account for another of these factors.

2 I know that leafy spurge can harm agricultural production. How would leafy spurge affect nonagricultural environments in eastern Idaho? I have been told that in Idaho, this plant grows mostly in humid conditions, and does not do well in arid conditions.

3 The table below lists four common methods used to control the spread of leafy spurge and some advantages and disadvantages of each method. Which one method do you recommend I use to control the leafy spurge on my property, and why?

4 Include a bibliographic list of all the references you used to answer questions 1–3.

METHOD	ADVANTAGES	DISADVANTAGES
Spray herbicide	• Works quickly • Repeated application may drastically reduce leafy spurge	• Expensive • May also harm crop plants or native plants • Must be applied at just the right time in the leafy spurge growth cycle
Beetles	• Free or nominal cost • Will not harm crop plants • Does not adversely affect the ecosystem	• Must be protected from insecticides, soil cultivation, burning, grazing, and so on for several years while they become established • Time intensive to collect • Can take five to ten years to become established and to have an effect • Will maintain leafy spurge at manageable levels but not eradicate
Sheep/Goats	• Leafy spurge is nontoxic and nutritious for sheep and goats • They prefer to eat flowering plants (like leafy spurge) and will leave the grasses to grow	• Must rotate animals among fields; it is costly to move them great distances • Cost of fencing • Need many animals to combat larger infestations • Seeds can get caught in hooves or hair and be transported to previously noninfested areas • Will maintain leafy spurge at manageable levels but not eradicate
Manual removal	• If done regularly for two to three years, may eradicate • No harm to other plants and grasses	• Plants must be pulled within seven days of germinating • If any of the root is left behind, it will grow again • Labor intensive • Only feasible in small, nonestablished infestations

HIGH SCHOOL ASSESSMENT RUBRIC

Use GIS to map a leafy spurge infestation and compute its area

STANDARD	EXEMPLARY	MASTERY	INTRODUCTORY	DOES NOT MEET REQUIREMENTS
4 The student knows and understands the changing physical characteristics of places.	Creates and prints a map layout that includes all of the required map elements (e.g., north arrow, legend, scale). The map accurately and clearly displays current and projected leafy spurge infestation.	Creates and prints a map layout that includes most of the required map elements (e.g., north arrow, legend, scale). The map accurately displays current and projected leafy spurge infestation.	Creates and prints a map layout that includes some of the required map elements (e.g., north arrow, legend, scale). The map attempts to display current and projected leafy spurge infestation, but is inaccurate in places.	Creates and prints a map layout that includes few of the required map elements (e.g., north arrow, legend, scale). The map displays either the current or projected leafy spurge infestation, but not both.
8 The student knows and understands the importance of ecosystems in people's understanding of environmental issues.	Describes how leafy spurge infestation could affect nonagricultural environments in eastern Idaho. Extensive research is evident by the many examples provided.	Describes how leafy spurge infestation could affect nonagricultural environments in eastern Idaho. Provides many examples.	Attempts to describe how leafy spurge infestation could affect nonagricultural environments in eastern Idaho. Provides few examples.	Does not attempt to describe how leafy spurge infestation could affect nonagricultural environments in eastern Idaho. Provides few examples.
14 The student knows and understands how to apply appropriate models and information to understand environmental problems.	Describes many different ways leafy spurge can spread. Includes a comprehensive list that covers spreading within the farmer's field, to other areas of the property, and to areas beyond the property. Presents and explains a simple model that incorporates one of the different ways leafy spurge is spread.	Describes different ways leafy spurge can spread. Includes a list that covers spreading within the farmer's field, to other areas of the property, and to areas beyond the property. Presents a simple model that incorporates one of the different ways leafy spurge is spread.	Describes one or two ways leafy spurge can spread. Attempts to present a simple model that incorporates one of the different ways leafy spurge is spread.	Describes one or two ways leafy spurge can spread, but does not incorporate them into a model.
18 The student knows and understands how to use geographic knowledge, skills, and perspectives to analyze problems and make decisions.	Creates accurate infestation projections and maps them. Uses the infestation projections to recommend an appropriate weed abatement method for the farmer. Explains how the recommended method will solve the farmer's leafy spurge infestation problem.	Creates accurate infestation projections and maps them. Uses the infestation projections to recommend an appropriate weed abatement method for the farmer.	Attempts to create infestation projections and maps them. Recommends a weed abatement method, but does not connect it to the farmer's situation.	Attempts to create infestation projections, but has difficulty mapping them. Recommends a weed abatement method, but does not connect it to the farmer's situation.

This is a four-point rubric based on the National Standards for Geographic Education. The "Mastery" level meets the target objective for grades 9–12.

On your own

Overview

This section provides guidelines and information to help you implement a similar project in your own community. This study gives students the opportunity to gather noxious weed data in the field and to witness how quickly weeds spread over time. The field experience and analysis of the maps they generate from the data can greatly enhance the students' understanding of the complexity and urgency of the noxious weed problem. A project like this allows them to use their knowledge and problem-solving skills to identify invasive plant "hot spots" and glimpse a future where invasive plants continue to flourish.

Estimated time

The length of time needed to complete this project can vary depending on the size of the area you will map. Shelley High began with a shorter project that was then extended to two years. If time is limited, you could choose to map a small study area at first.

- Start with a small, focused project that can be done in one to two weeks of class time.
- Use the small project to enable students to gather experience using the GPS units, field identification techniques, and the more advanced tools in ArcView.
- With the information and experience gained from completing the small project, you may wish to develop partnerships with local government or private agencies, agricultural or environmental groups, or a local farmer, and undertake a larger project.

Materials

- Computers for all students (one per student or student group)
- Access to a color printer or plotter to print out maps (optional)
- GPS units
- Street and aerial maps of the area you are going to study (By explaining that the data will be used for an educational project, you will often find that data providers are willing to waive fees for the data. Also, many states have clearinghouses with free data ready to use in ArcView.)
- Recording sheets for students to record the GPS and attribute data gathered in the field

Standards

GEOGRAPHY STANDARD	MIDDLE SCHOOL	HIGH SCHOOL
1 How to use maps and other geographic representations, tools, and technologies to acquire, process, and report information from a spatial perspective	The student knows and understands the relative advantages and disadvantages of using maps, globes, aerial and other photographs, satellite-produced images, and models to solve geographic problems.	The student knows how to use maps and other graphic representations to depict geographic problems.
3 How to analyze the spatial organization of people, places, and environments on Earth's surface	The student knows and understands how to use the elements of space to describe spatial patterns.	The student knows and understands how to use the models that describe patterns of spatial organization.
4 The physical and human characteristics of places	The student knows and understands how different human groups alter places in distinctive ways.	The student knows and understands the changing physical and human characteristics of places.
8 The characteristics and spatial distribution of ecosystems on Earth's surface	The student knows and understands how human activities influence changes in ecosystems.	The student knows and understands the importance of ecosystems in people's understanding of environmental issues.
14 How human actions modify the physical environment	The student knows and understands the consequences of human modification of the physical environment.	The student knows and understands how to apply appropriate models and information to understand environmental problems.
15 How physical systems affect human systems	The student knows and understands how the characteristics of different physical environments provide opportunities for or place constraints on human activities.	The student knows and understands the strategies to respond to constraints placed on human systems by the physical environment.
18 How to apply geography to interpret the present and plan for the future	The student knows and understands how to apply the geographic point of view to solve social and environmental problems by making geographically informed decisions.	The student knows and understands how to use geographic knowledge, skills, and perspectives to analyze problems and make decisions.

Objectives

The student is able to:

- Collect GPS points in an organized data table
- Create a map from the GPS data collected in the field
- Digitize polygons that represent weed infestations in the community
- Use ArcView tools to merge overlapping polygons and calculate the area in the desired units
- Predict future weed infestation and suggest an abatement plan
- Present the results of the study in a public forum
- Identify the difference between native and nonnative invasive species in your particular area

GIS skills and tools

The following table focuses on weeds as one type of invasive species. You can use similar procedures to study other invasive species (e.g., fire ants) if they are more relevant in your area.

GEOGRAPHIC INQUIRY STEP	GIS SKILL
Ask a geographic question	• Where are the weeds in your county now? • Can you tell by what means they are arriving and from where? • Does it appear that weeds are spreading to new areas along a source such as a waterway or road? • Is there a logical place to attack the noxious weed problem first that will stop the weed's spread to new areas?
Acquire geographic resources	• Contact local professionals to learn standards and requirements for the data to be collected in your area. • Form partnerships and obtain equipment, software, and other necessary support. Much of the data-gathering equipment can be expensive and many community entities will loan equipment and/or staff to help in your collections. • Practice using the GPS units in the field. • Ask an extension agent or your local garden store to help train you in weed identification and knowledge. • Ask a local agency or nature center, Bureau of Land Management office, state department of natural resources, or local farm cooperative to identify the areas that are priorities for mapping. • Systematically collect weed data in the field. Come up with a method of placing an imaginary grid on the field so that you do not miss any areas of noxious weed infestation.
Explore geographic data	• Download weed locations and attributes from the GPS data table. • Add the data table to ArcView. • Explore the patterns of weed locations. • Experiment with various orientation and interpretive themes to display with the weed data, roads, water bodies, aerial photographs, and parcel ownership maps.
Analyze geographic information	• Visually describe the overall distribution patterns for each species of weed. • Visually identify the areas with the greatest concentration of weeds. • Calculate the total infestation area. • Analyze changes in weed distribution over time by comparing data from one year to the next.
Act on geographic knowledge	• Provide local and state weed officials with hard-copy and electronic maps. • Prepare a slide show about the project and present the methods, findings, and benefits to professionals and public officials who are concerned with noxious weeds. • Create materials to educate the public about weeds such as identification keys, lists of poisonous plants, calendars, coloring books, and so on. • Partner with the local elementary school to share information about weeds and GIS. • Attend workshops and present findings at appropriate professional meetings and conferences. • Meet with local-, state-, and national-level government officials to relate the benefits of the project and discuss possible ways to expand it to other schools.

Teacher notes

Ask a geographic question

The following list represents several questions that can be addressed with a field study of invasive plants in your neighborhood.

• What plants are native to your area? Do a small research project with a local agency or nature center to learn about what has grown in your area in the past. Is there evidence of a native species being displaced by an aggressive non-native species?

• Are there areas around your school or community that are possible sites of nonnative weed infestation?

• Is there an area used for recreation that might be suffering the effects of human activity? For example, are there sensitive areas near a hiking trail or beach access where people wandering off the trail or taking shortcuts are damaging sensitive native plants? Is the disturbance allowing nonnative weeds or invasive species to become established?

• Are there agricultural fields in your area that may be affected?

• Do you have irrigation ditches or rivers and streams that could be access points for noxious weeds?

• Is there a county agency that has a project already started that could benefit from a partnership?

TEACHER TIPS

✓ Have a discussion with the students about their community and where invasive plants are growing.

✓ Contact a local extension agent to come into the class to discuss the invasive plant problems in your immediate area and get suggestions as to who might be willing to partner with students on this issue.

✓ Have the students brainstorm about local groups or individuals who would like to partner with the students on an invasive plant research project.

✓ Conduct preliminary research on noxious weeds in your area. Learn how to identify these weeds, how they spread, what damage they do, and how they reproduce. From your preliminary research, narrow your project to one or two weeds you have found present in your area.

✓ Find out the best time of year to collect data on invasive plants (What is the growing cycle? Is there a time that is not ideal for gathering data on the plants?).

✓ Plan your project with your partner(s). Include a projected time line and a list of immediate, short-term, and long-term goals.

✓ With a partner, determine a list of data attributes that the students will collect in the field. It is very important that students collect data that can be used by the appropriate state or county agencies for weed abatement.

Acquire geographic resources

TEACHER TIPS ON ACQUIRING BASEMAP DATA

✓ Local government or private agencies often store their GIS data in a map projection or coordinate system that is standard for their agency or your locale. When obtaining basemap data from these agencies, ask what their map projection standard is. Decide on a standard to use for your own project, and work with the agencies to acquire basemap data in that standard or one that you can easily convert.

✓ If you need to convert data from one map projection or coordinate system to another, ArcView includes a projection utility. Most GPS receivers allow you to choose the coordinate system in which locations are recorded. Use these tools to help your students make data in different map projections work together.

TEACHER TIPS ON PREPARING FOR FIELD DATA COLLECTION

✓ Get permission from landowners to enter private or government property to gather data. Visit each site in advance to see if there are hazards such as poisonous plants, unprotected ditches or other water hazards, snakes, or deep potholes that could injure students. Adjust site locations as necessary.

✓ Send home permission slips to parents that address the following: participation in the field data collection, student allergies, and the type of clothing and footwear necessary for field data collection.

✓ Find a partner who is willing to loan ten to twenty GPS units to your class for use in the field. That entity might be a local GPS distributor or GIS firm in your area and might wish to participate with you and your students in the field to lend technical expertise.

✓ Consult with a local expert to determine the list of attributes students need to collect while in the field.

✓ Obtain technical assistance in training your students to use the GPS units correctly and how to accurately and efficiently gather data in the field. Consult with your community partner for guidance.

✓ Divide the students into small data-gathering groups of two to three students. They should remain in their assigned groups throughout the project.

✓ Create a grid system and assign student groups to the squares on the grid. This method will allow you to keep track of where student groups are and to ensure that you have systematically collected data for the entire area.

✓ Design (or have students design) data-recording sheets to use in the field. A uniform data-recording sheet will help ensure consistency of the data collected.

✓ Conduct a practice data-gathering day on the grounds of your school. Have the students practice estimating distances and measurements of mock infestation areas. They should practice filling in the data-recording sheet. Demonstrate how to pace off an area and equate it to a measurement such as feet or acres. Make sure students practice recording GPS coordinates. They will need to understand how to record a specific location on the GPS unit.

TEACHER TIPS ON FIELD DATA COLLECTION

✓ Arrange for student transportation in advance.

✓ Monitor the weather. Find out if there are some conditions that may not be optimal for walking through fields of noxious weeds (e.g., very windy conditions, rainy weather).

✓ Ask other teachers if they will chaperone during the field data collection. If they do attend, be sure to assign them a specific task (e.g., help the students identify plants or techniques for estimating sizes of infestation or fields).

✓ Invite parents to go on the field data collection day as chaperones. Assign parents a specific task such as working with two or three student groups or checking equipment in and out to all student groups.

✓ Before you leave the data collection site, collect all data sheets from each student group.

Explore geographic data

Technical issues

When collecting field data, there are a number of technical issues you may encounter. Here are some tips for dealing with them.

- When you return to the classroom, you will have multiple data-recording sheets from your student groups and it will be necessary to aggregate all the information into one form. Have each group create a spreadsheet with their data, and then merge the spreadsheets into a single file.
- Create a template spreadsheet and give it to all the students to fill out. This table should have the same structure that was used to collect data in the field. Make sure your field headings do not have any spaces in them and avoid using symbols such as the "%" sign.
- Remind students which numbers should be latitude (northing) and which should be longitude (easting). If they are recording latitude and longitude coordinates and are working in the western or southern hemisphere, some of the GPS coordinates may need a minus sign in front.

Sample data table

SITE_ID	GPS_DATE	LATITUDE (NORTHING)	LONGITUDE (EASTING)	SPECIES	PHENOLOGY	SIZEINFEST (IN ACRES)	PERCENT_ COVER

- Have the students save their completed spreadsheets on a floppy disk or in a designated folder on the school network.
- Combine all the students' tables into one by copying and pasting their data into a master spreadsheet.
- Save this spreadsheet as a .dbf or .txt file.
- Do a dry run with the spreadsheet to be sure you can load it into ArcView and the points plot on a map correctly. If it does not work properly, you should make adjustments to the spreadsheet prior to having the students attempt to work with it in ArcView.
  - Remember, ArcView permits different functions on numeric and string (text) fields. If you have numeric data, make sure those spreadsheet cells are formatted as number and not text.
  - If only part of your data loaded into ArcView, check to be sure that no cells are selected when you export the spreadsheet.
  - If you have saved the file in .dbf format and are having problems bringing it into ArcView, try saving the file as tab-delimited text instead.
- Save the tested master spreadsheet with the student groups' data onto a floppy disk for each student or save it in a designated folder on the school network.
- If you have downloaded or acquired additional data such as digital orthophoto quadrangles, copy that data to the same place where you put the master data spreadsheet. Keep all of the project data together.

Preliminary data exploration

- Have the students look at the completed master table in a spreadsheet program. Ask why the data looks the way it does. For example, are the latitude and longitude readings consistent? Look at the tenth- and hundredth-place digits of the latitude and longitude data to see whether they appear to be correct. This is a good way to find simple transposition or copy errors. In the following example, the third longitude value is obviously inconsistent with the others.

LONGITUDE
−118.15929
−118.15935
−118.35940

You will need to consider the size of the study area in determining how "consistent" the values should be. A difference of one tenth of a degree represents a difference of up to 7 miles on the ground, so if the study area is less than 7 miles across, the latitude/longitude values should all be within one tenth of a degree.

- Assign students the task of evaluating the data their group collected in the field. They should check to make sure the infestation size represented in the map coincides with what they collected. If a student group notices an error, they need to inform the entire class.

- Symbolize the data.
- What visual patterns do students notice with the data? For example, was only one species observed or were other species represented in the field as well?
- Students may need to perform some additional steps to map their field data. For example, if they collected point locations, they will need to add an event theme to their view. If they collected locations defining the perimeter of infestation areas, the method for adding polygons to the view will depend on the type of GPS unit they use. For example, in some cases the GPS unit can generate the polygons. In other cases, they may need to run a script, such as the sample script GPS2shp.avl that comes with ArcView.
- Students can digitize circles around the points of infestation as illustrated in the exercise. If students are going to create circles (or other shapes) around point locations, as in the exercise, help them develop a simple model or guidelines for determining the radius or other dimension to use based on the attribute values or classes.

  Note for ArcView 3.1–3.3 users: You can have the students use the Geoprocessing Wizard to buffer the points and merge the shapes automatically.

Analyze geographic information
Analyzing infestation areas

- Instruct students to merge and calculate their completed circles or polygons so that they can get the total square feet of the infestation.
- Compute the percent of coverage of the invasive plants in the field by dividing the infestation area by the total area of the field and multiplying by 100.
- Look at the map to see if there seems to be heavier infestation closer to certain physical features (e.g., roads, drainage ditches, streams and rivers, center of fields, edges of fields). Students may want to calculate different infestation percentages for different areas if they notice a pattern.

Drawing conclusions

- Hypothesize where the invasive plant infestations could be coming from (e.g., if they are lining roadways, possibly they are being transported by farm equipment or vehicles). You can have your students analyze an aerial photograph or satellite image more critically to see if there is another infestation on the map that is connected with their fields by a road or if the infestation runs along the border of a particular road and there is a similar infestation at both ends of that road.
- Have the students predict what the infestation might look like in two to five years. Without having the students go through the whole process of digitizing more circles and merging and calculating areas again, ask the students to estimate how the infestations will increase in size in the future. In this module's assessment, the students were asked to project leafy spurge infestation after four years. Your students can use a similar model to extrapolate how quickly the infestations will spread in the future.
- Research possible weed abatement methods useful in your area. Develop a basic abatement plan that is based on research, infestation percentage, concentration of the plants, and speed at which the invasive plants propagate.

Act on geographic knowledge
When we drive along a road or go hiking along a bike path we are usually unaware of what invasive species may be present or that we may be inadvertently spreading such species. Increasing the level of public awareness about invasive plants is essential to controlling them. Having your students share their work with the public is important so that the whole community can benefit from what they learned about noxious weeds.

- Share your students' results with the owner of the land where the data was collected.
- Develop a public awareness campaign with poster-sized maps, leaflets, and flyers. Advise the public about invasive weeds in the area and what they can do to avoid spreading them.
- Do presentations in elementary schools to educate younger students on what invasive species are and how they can help in the campaign against noxious weeds in their area.
- Have the students arrange to be on the agenda for a town council meeting where they use multimedia and layouts to present their study, findings, and possible recommendations.
- Invite the local media to produce a story on the students' work and their results and suggestions. This may be a springboard to internships or paid after-school and summer work for your students.

Module 4: Tracking water quality

Lesson

Case study: Monitoring seasonal changes on the Turtle River
Exercise: Analyze Turtle River data to identify locations for fish habitat restoration

Overview

Students will explore a small river and its chemical makeup. They will analyze chemical properties in water samples that have been collected at five sites along the river and will determine which sites are appropriate for fish habitat restoration.

Estimated time

45-MINUTE CLASS	ACTIVITY
1	Lesson introduction
2	Exercise part 1
3	Exercise part 2
4	Assessment

Materials

- *Community Geography: GIS in Action* book (one per student)
- Computers for all students (for the exercise and assessment)
- Color printer (to print the student maps)

Student handouts from this lesson to be photocopied:

- Assessment

Standards

GEOGRAPHY STANDARD	MIDDLE SCHOOL	HIGH SCHOOL
3 How to analyze the spatial organization of people, places, and environments on Earth's surface	The student knows and understands how to use the elements of space to describe spatial patterns.	The student knows and understands how to apply concepts and models of spatial organization to make decisions.
8 The characteristics and spatial distribution of ecosystems on Earth's surface	The student knows and understands how physical processes produce changes in ecosystems.	The student knows and understands the importance of ecosystems in people's understanding of environmental issues.
18 How to apply geography to interpret the present and plan for the future	The student knows and understands how to apply the geographic point of view to solve social and environmental problems by making geographically informed decisions.	The student knows and understands how to use geographic knowledge, skills, and perspectives to analyze problems and make decisions.

Objectives

The student is able to:

- Thematically map data using graduated symbols and bar charts
- Add an event theme
- Join new tables to existing maps
- Temporally and spatially analyze data to identify locations for fish habitat restoration

GIS skills and tools

📷	Use the Theme Properties button to rename a theme	ⓘ	Use the Identify tool to learn more about a selected record
🔢	Open the attribute table for a theme	⊞	Join tables
✏️	Zoom in on the active theme	▶	Select and move labels
🔍	Zoom in on the view	▣	Clear selected features

GEOGRAPHIC INQUIRY STEP	GIS SKILL
Ask a geographic question	• Recognize and understand geographic questions posed in a scenario.
Acquire geographic resources	• Create a basemap by adding state, watershed, and river themes. • Add tabular water-testing data to the project. • Use the Add Event Theme function to create a new shapefile of data.
Explore geographic data	• Observe the spatial and temporal relationship of water quality variables by: 　– Turning themes on and off 　– Copying and pasting themes 　– Changing theme names 　– Joining new season tabular data to an existing table • Explore attributes using the Identify tool and attribute table.
Analyze geographic information	• Classify water quality data by variable and season. • Symbolize data with bar charts. • Spatially analyze the data to determine which site is suitable for fish habitat restoration.
Act on geographic knowledge	• Recommend one site for fish habitat restoration.

Teacher notes

Lesson introduction

In this exercise your students will explore the condition of the Turtle River in North Dakota. This river has no startling environmental problem, but it does illustrate how the geology of the area and the change of seasons affect the river's chemical makeup.

It would be best to introduce this lesson after the students have completed a lesson or unit on chemical solutions so they understand trends in solubility. It would also be appropriate to do this lesson at the end of a biology lesson or unit on wetland ecosystems. It is not imperative to introduce these concepts first, but if the students have not been introduced to them at all, it is recommended that you spend a class period on the introductory activity described below. The introductory activity will help your students acquire basic knowledge about water-quality testing.

Introductory activity

Divide the students into groups of three or four and assign each group to research one of the following topics:

- Water quality—what is it and how is it defined?
- The importance of water to living organisms
- Aquifers
- pH
- The effects of water temperature
- Dissolved oxygen
- Conductivity

Instruct each group to conduct their research at one of the suggested Web sites listed below. Each group should prepare to deliver a five-minute presentation on their topic to the class. They can use a whiteboard or overhead projector to display information. Remind students to use more than one Web site in their research and to cite their sources during the presentations.

WEB SITE NAME	WEB SITE ADDRESS	SEARCH TIPS
The GLOBE Program	www.globe.gov	Search the GLOBE Web site for keywords such as hydrology, water quality, pH, aquifers, dissolved oxygen, conductivity, and temperature.
River Network	www.rivernetwork.org	Search the River Network Research Library for keywords such as watershed health, water quality, and pH.
USGS Water Science for Schools	ga.water.usgs.gov/edu	Explore Earth's Water for information on the importance of water and aquifers. Explore Water Basics for information on pH, dissolved oxygen, temperature, and conductivity.
EPA Environmental Education Center	www.epa.gov/teachers and www.epa.gov/students	On the teacher's site, explore Background Information on water. On the student's site, explore the water section for your assigned topic.

Exercise

Before completing this lesson with students, we recommend that you complete it as well. Doing so will allow you to modify the activity to accommodate the specific needs of your students.

Read the exercise scenario aloud with the class. Explain to the students that in part 1 they will be studying the watershed and aquifers in the area of the Turtle River, as well as the river itself. They will add a table to ArcView and symbolize one characteristic of the river (depth).

In part 2 the students will join a fourth season's data table to the original table used in part 1, symbolize three more characteristics, and finally analyze all four seasons' worth of data.

TEACHER TIPS

✓ Part 1 is shorter than part 2. If your students have time in one class period, they can progress to part 2. Remind them to save their work when the class period ends.

✓ Be sure to give students instructions on how to rename the project, save it, and record the new name so they can open it during another class or from a different computer.

Things to look for while students are working on this activity:

• Are the students thoughtfully answering the exercise questions?
• Are the students able to successfully add a table and join a table?
• Does each student end up with four new themes added to their view (one for each variable)?
• Students should fill out the table for each variable as they go, and not wait until the end of the exercise to fill them all out.
• Students should be able to go back to each completed table to find the information they need for the final table and to decide where the fish habitat should be located.

Conclusion

Conclude the lesson by having a class discussion centered about the following questions:

• Which sites did the students determine were suitable for the fish habitat restoration project?
• Why were these sites chosen?
• Where might other applications of these techniques be used and why?
• What can happen if pH in the river changes by 1?

Assessment

For both the middle school and high school assessments, students will be given information about what conditions are desirable for supporting macroinvertebrates (the aquatic insects, worms, and crustaceans that live in rivers). Students will use the criteria and the project and data from the exercise to identify locations and seasons that would be appropriate for creating human-built habitat for macroinvertebrates.

Middle school: Highlights skills appropriate to grades 5 through 8

The students will be divided into teams and assigned a particular site. Each team will evaluate one site's set of criteria describing ideal conditions for macroinvertebrates and, in the large classroom group, will determine at which site and during which season structures should be built to provide macroinvertebrate habitat. They will create maps of their site's data, present to the class, and write a paragraph explaining their conclusions.

High school: Highlights skills appropriate to grades 9 through 12

The students will be divided into teams. Each team will evaluate a set of criteria for the five sites describing ideal conditions for macroinvertebrates. They will determine at which site and during which season structures should be built to provide macroinvertebrate habitat. They will create maps of all the data and write an essay explaining their conclusions.

MIDDLE SCHOOL ASSESSMENT

Analyze Turtle River data to identify locations for fish habitat restoration

Name _____   Date _____

Assigned site _____

A nature conservancy group has asked your class to help them with a new project using the existing data you have on the Turtle River. Fish habitats have been restored and fish are becoming more plentiful in the river. The nature conservancy would like to improve the food sources for the fish, which eat macroinvertebrates such as aquatic insects, worms, and crustaceans. They plan to do this by building structures that provide the habitat macroinvertebrates need to reproduce and survive. The conservancy needs you to evaluate the data and determine which sites meet the criteria listed below for macroinvertebrate structures and which season would be the best to build these structures.

SITE SELECTION CRITERIA	SEASON SELECTION CRITERIA
• pH must be in the range of 7.0–8.49. • Dissolved oxygen levels must be 10.0 ppm or greater.	• Temperature must be 15 degrees centigrade or higher. • Structure needs to be built at least three months before the river freezes solid, in order to give the macroinvertebrates enough time to reproduce.

Perform the following steps in order

1 Use the project and data from the exercise to complete the data table for your assigned site. Use a separate sheet of paper and use a highlighter to highlight each variable that meets the above-stated criteria.

Your assigned site _____

VARIABLE	SUMMER	FALL	SPRING
Dissolved oxygen			
pH			
Temperature			

2 Create a presentation layout of your assigned site's map results, showing dissolved oxygen, pH, and temperature.
- Create a different map for each variable and display the three maps on one layout.
- Be sure to include appropriate symbology, your group members' names, date, title, scale bar, and north arrow.

3 Present your group's site information to your class.
- Explain the data for your site.
- Determine whether your site is appropriate for macroinvertebrate structures.

4 Write a paragraph explaining your group's conclusions.
- After discussing all the site data with your class, determine which sites and which seasons are appropriate for building macroinvertebrate structures.
- Which sites and seasons were optimal and why?

MIDDLE SCHOOL ASSESSMENT RUBRIC

Analyze Turtle River data to identify locations for fish habitat restoration

STANDARD	EXEMPLARY	MASTERY	INTRODUCTORY	DOES NOT MEET REQUIREMENTS
3 The student knows and understands how to use the elements of space to describe spatial patterns.	Creates and compares maps of variables across seasons to determine which site is best for building macroinvertebrate structures. The maps contain appropriate symbology, labeling, name, date, and legend.	Creates a layout of maps comparing three variables across seasons at their assigned site. The maps contain appropriate symbology and labeling.	Creates a layout of maps comparing two of the variables across seasons at their assigned site. The maps contain symbology and labeling, but some are incorrect.	Creates a map of one variable across seasons at their assigned site. The maps are missing most symbology and labeling.
8 The student knows and understands how physical processes produce changes in ecosystems.	Correctly identifies which variables and season meet the macroinvertebrate requirements at the assigned site. Evaluates all the site data and correctly selects one that best fits the macroinvertebrate ecosystem requirements.	Identifies which variables and season meet the macroinvertebrate requirements at the assigned site. Evaluates most of the site data to select one that best fits the macroinvertebrate ecosystem requirements.	Incorrectly identifies which variables and season meet the macroinvertebrate requirements at the assigned site. Attempts to evaluate the site data to select one that best fits the macroinvertebrate ecosystem requirements.	Has difficulty identifying which variables and season meet the macroinvertebrate requirements at the assigned site. Does not select a site that best fits the macroinvertebrate ecosystem requirements.
18 The student knows and understands how to apply the geographic point of view to solve social and environmental problems by making geographically informed decisions.	Presents a clear argument as to why their solution is the most logical based on the analysis of the data. Is able to articulate the process they went through in developing the solution.	Presents a clear argument as to why the solution is the most logical based on the analysis of the data.	Presents a solution to the problem, but provides little evidence for the solution.	Presents a solution to the problem, but provides no evidence for the solution.

This is a four-point rubric based on the National Standards for Geographic Education. The "Mastery" level meets the target objective for grades 5–8.

HIGH SCHOOL ASSESSMENT

Analyze Turtle River data to identify locations for fish habitat restoration

Name _____ Date _____

A nature conservancy group has asked your class to help them with a new project using the existing data you have on the Turtle River. Fish habitats have been restored and fish are becoming more plentiful in the river. The nature conservancy would like to improve the food sources for the fish, which eat macroinvertebrates such as aquatic insects, worms, and crustaceans. They plan to do this by building structures that provide the habitat macroinvertebrates need to reproduce and survive. The conservancy needs you to evaluate the data and determine which sites meet the criteria listed below for macroinvertebrate structures and which season would be the best to build these structures.

SITE SELECTION CRITERIA	SEASON SELECTION CRITERIA
• pH must be in the range of 7.0–8.49. • Dissolved oxygen levels must be 10.0 ppm or greater.	• Temperature must be 15 degrees centigrade or higher. • Structure needs to be built at least three months before the river freezes solid, in order to give the macroinvertebrates enough time to reproduce.

Perform the following steps in order

1 Use the project and data from the exercise to determine which of the five sites you are mapping meet all the criteria. For each such site, complete a data table similar to the one below:

Site _____

VARIABLE	SUMMER	FALL	SPRING
Dissolved oxygen			
pH			
Temperature			

2 Create a presentation layout of your map results showing dissolved oxygen, pH, and temperature.
  • Create a different map for each variable and display the three maps on one layout.
  • Be sure to include appropriate symbology, your name, date, title, scale bar, and north arrow.

3 Write an essay explaining your conclusions.
  • Describe which one or more sites and seasons meet the macroinvertebrate structure criteria and why they were selected.

HIGH SCHOOL ASSESSMENT RUBRIC

Analyze Turtle River data to identify locations for fish habitat restoration

STANDARD	EXEMPLARY	MASTERY	INTRODUCTORY	DOES NOT MEET REQUIREMENTS
3 The student knows and understands how to apply concepts and models of spatial organization to make decisions.	Creates three maps that compare the data temporally (across seasons) and spatially (across space). The maps contain appropriate symbology, labeling, name, date, and legend. The maps are used to determine sites that meet the macroinvertebrate structure criteria.	Creates three maps that compare the data temporally and spatially. The maps contain appropriate symbology and labeling. The maps are used to determine most of the sites that meet the macroinvertebrate structure criteria.	Creates maps that attempt to compare the data temporally and spatially. The maps contain symbology and labeling, but some is incorrect. Has difficulty determining which sites meet the macroinvertebrate structure criteria.	Creates maps, but they do not compare the data temporally and spatially. The maps contain symbology and labeling, but most is incorrect. Is not able to determine which sites meet the macroinvertebrate structure criteria.
8 The student knows and understands the importance of ecosystems in people's understanding of environmental issues.	Correctly identifies how each variable (temperature, pH, and dissolved oxygen) affects the river's ecosystem at all sites and selects a site that best fits the macroinvertebrate ecosystem requirements.	Correctly identifies how each variable affects the river's ecosystem at most sites and selects a site that best fits the macroinvertebrate ecosystem requirements.	Correctly identifies how some variables affect the river's ecosystem at most sites and has difficulty selecting a site that best fits the macroinvertebrate ecosystem requirements.	Incorrectly identifies how some variables affect the river's ecosystem at some sites and does not select a site that best fits the macroinvertebrate ecosystem requirements.
18 The student knows and understands how to use geographic knowledge, skills, and perspectives to analyze problems and make decisions.	Presents a clear written argument as to why their solution is the most logical. Is able to articulate the geographic inquiry process they went through to develop the solution.	Presents a clear written argument as to why their solution is the most logical. Is able to articulate most of the geographic inquiry process they went through to develop the solution.	Presents a written argument as to why their solution is the most logical. Is unable to articulate the geographic inquiry process they went through to develop the solution.	Does not present a written argument as to why their solution is the most logical. Is unable to articulate the geographic inquiry process they went through to develop the solution.

This is a four-point rubric based on the National Standards for Geographic Education. The "Mastery" level meets the target objective for grades 9–12.

On your own

Overview

This section provides guidelines and information to help you implement a similar project in your own classroom. Studying a river, creek, or any environmental phenomenon over a period of time provides a unique opportunity for your students to analyze trends and to witness change over time. The students will need to develop and draw upon problem-solving skills: how to collect and organize data, how to integrate information from various sources, how to analyze patterns and relationships within data, how to collaborate with others in decision making, and how to communicate results in a variety of formats.

Estimated time

- One school day for each water-sampling session. (This can be adjusted depending on how many sites you intend to study. One site might only take one class period, whereas four or five sites will take a whole school day.)
- One or two class periods for mapping and analyzing data for each day of data collected in the field.

Materials

- Computers for all students (one per student or student group)
- Reference materials on water-quality testing in schools (see the module 4 resources and references section)
- Water-testing equipment (comes in varying complexity; choose products that suit your budget and student audience; refer to *Community Geography: GIS in Action,* module 4, "On your own," for a list of water-testing materials)
- Data sheet to remind students which variables they are collecting and to provide consistency in your results (you need to create this sheet)
- Class e-mail address for correspondence with community partners
- Color printer

Standards

	GEOGRAPHY STANDARD	MIDDLE SCHOOL	HIGH SCHOOL
1	How to use maps and other geographic representations, tools, and technologies to acquire, process, and report information from a spatial perspective	The student knows and understands how to make and use maps, globes, graphs, charts, models, and databases to analyze spatial distributions and patterns.	The student knows and understands how to use geographic representations and tools to analyze, explain, and solve geographic problems.
3	How to analyze the spatial organization of people, places, and environments on Earth's surface	The student knows and understands how to use the elements of space to describe spatial patterns.	The student knows and understands how to apply concepts and models of spatial organization to make decisions.
4	The physical and human characteristics of places	The student knows and understands how different physical processes shape places.	The student knows and understands the changing physical and human characteristics of places.
7	The physical processes that shape the patterns of Earth's surface	The student knows and understands how physical processes shape patterns in the physical environment.	The student knows and understands the interaction of Earth's physical systems.
8	The characteristics and spatial distribution of ecosystems on Earth's surface	The student knows and understands how physical processes produce changes in ecosystems.	The student knows and understands the importance of ecosystems in people's understanding of environmental issues.
14	How human actions modify the physical environment	The student knows and understands how human modifications of the physical environment in one place often lead to changes in other places.	The student knows and understands the significance of the global effects of human modification of the physical environment.
15	How physical systems affect human systems	The student knows and understands how the characteristics of different physical environments provide opportunities for or place constraints on human activities.	The student knows and understands strategies to respond to constraints placed on human systems by the physical environment.
18	How to apply geography to interpret the present and plan for the future	The student knows and understands how to apply the geographic point of view to solve social and environmental problems by making geographically informed decisions.	The student knows and understands how to use geographic knowledge, skills, and perspectives to analyze problems and make decisions.

Objectives

The student is able to:

- Collect and label water samples with location, time, and date
- Take GPS coordinates
- Record and compile data into a spreadsheet
- Analyze the sample data using GIS
- Create a map or layout illustrating the results of their study
- Present the results to the public through a variety of channels

GIS skills and tools

The following table focuses on weeds as one type of invasive species. You can use similar procedures to study other invasive species (e.g., fire ants) if they are more relevant in your area.

GEOGRAPHIC INQUIRY STEP	GIS SKILL
Ask a geographic question	• Where is an accessible and interesting river in your area? • What are local water-related issues that can be studied? • What chemicals are dissolved in the river and does their concentration vary from place to place? • What are the environmental demands on the river?
Acquire geographic resources	• Work with local organizations willing to provide resources for your project. • Work with local government agencies that have an interest in your geographic region (e.g., water and soil extension agents, parks and recreation employees, state and national wildlife refuge employees). • Work with local professionals in water data collection to plan your study and analysis. • Collect water samples at four to eight different sites along the river.
Explore geographic data	• Sort your data in the attribute table. • Join new data to previous data sets. • Thematically map different attributes. • Observe spatial patterns of water-quality data using appropriate symbology.
Analyze geographic information	• Visually analyze spatial patterns of water-quality data. • Use theme-on-theme selection to find areas of environmental concern.
Act on geographic knowledge	• Create and print a layout to display the results of spatial analysis. • Create signs and brochures to educate the public about your results. • Make presentations to local government and various community organizations.

Teacher notes

Ask a geographic question

To start a water study project, you must help your students develop an affinity for the project at hand. If you are going to conduct a water study with your students, you must have a river, pond, or creek somewhere in the vicinity. Chances are your students have had some interaction with the body of water whether in a positive or negative way. As a precursor to your study, lead students on an activity that enhances their sense of place around water. Use the following suggestions:

- Have students write about a memory they have that is centered around water (e.g., fishing trip, picnic, summer camp, boat ride).
- Have students read text or poetry set around water and discuss the piece as a class (e.g., Mark Twain's novel *The Adventures of Huckleberry Finn* and Ernest Thompson's play *On Golden Pond*).
- Go on a field trip to a nearby river or creek and assign students to create a pictorial survey of the area using cameras, their artwork, and their words. Students can record animals, plants, and insects.

Once you have students thinking about water in their environment, they should have natural curiosity about what lives in and around your local waterways. As a way to focus your students on selecting a specific research topic, explore the following questions:

- Are there local waterways that the public perceives as being polluted?
- What water-related issues have been in the news recently?
- Is there a waterway near your school? Are students interested in evaluating the health of that waterway?
- Was there a recent flooding event that has captured local attention?

> **TEACHER TIPS ON SITE SELECTION**
>
> ✓ In addition to the site selection tips outlined in *Community Geography: GIS in Action,* you will need to take the age of your students into consideration when choosing a study site.
> ✓ For younger students, choose sites closer to school.
>   – Choose streams or creeks that do not pose major downstream hazards.
>   – Choose streams or creeks that are no deeper than 2 feet and are not fast moving.
> ✓ For older students, choose sites no more than a half-hour drive from the school.
>   – Choose a small river or another body of water with water depth less than 5 feet.
>   – Choose streams or creeks that do not pose major downstream hazards.
>
> If local water sources are not easily accessible, consider modifying the project to accommodate your local surroundings. For example, you can collect different data and analyze it accordingly. Here are some possibilities:
>
> | | | |
> |---|---|---|
> | Weeds | Trees | Animal movement |
> | Animals | Plants | Rain amount over time |
> | Insects | Litter | Water and air temperature |

Acquire geographic resources

> **TEACHER TIPS ON PREPARING FOR FIELD DATA COLLECTION**
>
> ✓ Send home permission slips to parents that cover the following: participation in the field testing, the child's swimming ability, and whether they prefer that their child stay out of the water.
> ✓ In the case of younger students or a large number of students who cannot swim, have each student wear a life jacket.
> ✓ Arrange for a professional chemist to visit your class to demonstrate testing techniques and to share with your students what is done at their lab.
> ✓ Arrange for a community geologist to come in and do a grade-level-appropriate geology lesson. The lesson can include local glacial activity, underground aquifers, groundwater and surface water issues, and geologic formations present in the area. Make sure you inform the professional of your students' grade level and the type of room he or she will be presenting in. Help the geologist prepare by sharing ideas for hands-on activities that will engage the students and imprint the content presented.
> ✓ In advance, divide the students into testing groups. They will remain in these groups throughout the project.
> ✓ Have each testing group practice the testing techniques. Bring in bottles of water from different parts of the town (make sure to refrigerate them) for the students to use in practicing their testing procedures. In this way, the students will become proficient with unfamiliar equipment and testing kits prior to the day in the field.
> ✓ Have a separate practice day where each testing group learns how to take and record GPS coordinates.
> ✓ Design data-recording sheets for your students to use in the field or have students design these in advance. A uniform data-recording sheet will help to maintain consistency with all the data collected.
> ✓ Make sure you have permission to access each site. Visit each site in advance to see if there are any possible hazards such as poisonous plants or deep potholes in the river. Adjust sites as necessary.

Explore geographic data

Technical issues

- When you return to the classroom you will have multiple data-recording sheets and it will be necessary to aggregate that information into one form for students to enter. Set up a data table on the whiteboard or an overhead projector, where you or a student records the data for each site and variable you tested. Invite each testing group to record their data and discuss their results with the class. This process is relatively quick and should take only fifteen or twenty minutes while providing time for the students to get a first look at the aggregated data. They will also spot immediately any erroneous data and can troubleshoot what might have happened (e.g., incorrect decimal place).

Sample data table:

SITE #	LATITUDE	LONGITUDE	TEMP	DEPTH	pH	DO
Site 1						
Site 2						
Site 3						

If you have several student "experts" in each class with data for the same site, you can adjust your table to allow for averaging of each variable:

Site 1

GROUP #	LATITUDE	LONGITUDE	TEMP	DEPTH	pH	DO
Expert 1						
Expert 2						
Average	**********	**********				

Site 2

GROUP #	LATITUDE	LONGITUDE	TEMP	DEPTH	pH	DO
Expert 1						
Expert 2						
Average	**********	**********				

- Have students check their latitude and longitude data to make sure they do not have a number that is way out of line with the others. Remind them that even if the sample sites were 5 to 10 miles apart, only the last three decimals will change and they will be fairly close together numerically.
- Remind students that they do not need to do any special formatting in their tables such as changing fonts or centering the text when they are creating their spreadsheets.
- Have students enter the aggregated averaged data in a spreadsheet. If you have a template set up for your students to use, they should be able to enter the spreadsheet in the balance of the class period in which they aggregated the data on the board.
- If you have downloaded additional data for the students to use (e.g., aquifer data from the Internet), instruct students as to where they can locate these files.
- After students add their spreadsheet to their project, you will most likely have to "debug" some of the tables. If students have problems getting the table to map, check the following:
  - Are the latitude and longitude fields formatted with the same number of decimal places your GPS unit recorded? If not, ArcView might have rounded the values to two decimal places and all the sample sites will be in the same place.
  - For western or southern hemisphere locations, are the latitude and longitude coordinates negative? If not, you need to edit the table appropriately. Once these changes are made, the coordinates should map correctly.
  - Are all the numeric fields formatted as numeric and not text? If they are not, you will not be able to thematically map the numeric data with any legend type but Unique Value. You can tell if a field is numeric if data is right justified in the table when you bring it into ArcView.

Preliminary data exploration

Your preliminary data exploration starts when you are aggregating the data collected by each group of students. While recording student data in front of the class, ask the students to examine the data:

- Does the data differ significantly between the groups who collected the data?
- Do the students feel it would be beneficial to replace the individual data sets with averages from multiple groups?
  - If you choose to average, the data will most likely not be usable by an outside or professional group.
  - If you choose *not* to average, you will have to figure out a way to choose which groups' data to use in the analysis.
  - The students can give you feedback as to how well their data-gathering efforts went, which might help you ascertain which data has more integrity.
- Ask questions as to why the data looks the way it does. For example,
  - Regarding conductivity, was there a rainstorm in the last day or two?
  - If the nitrate levels are really high, did they see a farmer plowing or fertilizing the field next to the river?
  - In the case of temperature, was there a recent cold snap or heat spell?
  - Is there something else in the environment that could have affected their data?

Analyze geographic information

- Look for any patterns in the water-quality data such as values outside the normal range, for example very high or low pH (< 6 and > 8), low dissolved oxygen (under 6 ppm), or fecal coliform (numerous colonies).

  - Look for geologic patterns in your data. For instance, the National Atlas *(www.nationalatlas.gov)* contains all the principal aquifers in the United States so that you can look to see if the chemistry corresponds to where the aquifers are. Geological maps may be helpful but are sparse on the Internet because not as many have been digitized.

  - Next, look for possible relationships between land use and patterns in the water-quality data. For example, students can examine aerial photographs to look for large lawns or agricultural areas; runoff from these areas can cause high levels of nitrates, especially if they have been fertilized recently. In addition, livestock grazing near waterways might increase the number of fecal coliform colonies. Also, elevated levels of fecal coliform could be caused by discharge from nearby sewage treatment ponds. Conditions like high turbidity may be caused by recent forest fires, which can cause high soil runoff.

- Once you see what the patterns of the chemical concentrations are and how they change seasonally, see if these patterns are consistent spatially.

Act on geographic knowledge

Waterways are sometimes overlooked and your students' work can help the public appreciate their community water sources in a way they never before have. Consider the following projects as gateways to community outreach:

- Share your students' results with the city parks program and suggest a campaign to put up interpretive signs along the river showing the results of the students' research. This can be the springboard for offering good advice to the community, such as encouraging people to fertilize wisely, or organizing environmental clubs to do litter pickups, or advising people as to what happens to their car oil if they discard it in the gutter.

- Lobby local parks associations for developing recreation trails along waterways.

- Have the students organize a communitywide "Take Your Kid Fishing Weekend" to instill in town residents an appreciation of their local aquatic resource.

- Have the students arrange to be on the agenda for a town council meeting where they use multimedia and layouts to present their study, findings, and possible recommendations.

- Share the project findings with other classes in the school as well as other schools in the district. Have the students teach classes about their local river.

- Join with other surrounding schools that may test water in a different part of the same river system and evaluate the water from a larger perspective.

Lesson

Case study: Identifying potentially harmfull landfills
Exercise: Map, query, and analyze neighborhood data to identify high-risk landfills

Overview

In this lesson, students will explore the means by which human societies dispose of their wastes and the potential risks of these waste-disposal strategies. They will use GIS to map and analyze one such example—abandoned landfill sites in a Toronto neighborhood. Their task will be to identify the abandoned landfill sites that represent the greatest threat to the people and natural resources in that neighborhood. In part 1 of the exercise, students will create a basemap reflecting characteristics of the neighborhood by adding and symbolizing cultural and physical features themes. In part 2, students will add and explore landfill site data. Finally, they will identify and map high-risk landfills in part 3.

Estimated time

45-MINUTE CLASS	ACTIVITY
1	Lesson introduction and exercise part 1
2	Exercise parts 2 and 3
3	Conclusion and begin assessment
4	Complete assessment

Materials

- *Community Geography: GIS in Action* book (one per student)
- Computers for all students (for the exercise and assessment)
- Transparency: Trends in waste generation, recovery, and disposal

Student handouts from this lesson to be photocopied:

- Assessment

Standards

GEOGRAPHY STANDARD	MIDDLE SCHOOL	HIGH SCHOOL
1 How to use maps and other geographic representations, tools, and technologies to acquire, process, and report information from a spatial perspective	The student knows and understands how to make and use maps and databases to analyze spatial distributions and patterns.	The student knows and understands how to use geographic representations and tools to analyze, explain, and solve geographic problems.
4 The physical and human characteristics of places	The student knows and understands how to analyze the human characteristics of places.	The student knows and understands the changing physical and human characteristics of places.
14 How human actions modify the physical environment	The student knows and understands the consequences of human modification of the physical environment.	The student knows and understands how to apply appropriate models and information to understand environmental problems.
18 How to apply geography to interpret the present and plan for the future	The student knows and understands how to apply the geographic point of view to solve social and environmental problems by making geographically informed decisions.	The student knows and understands how to use geographic knowledge, skills, and perspectives to analyze problems and make decisions.

Objectives

The student is able to:

- Compare and contrast the use of landfills for waste management in the late twentieth century with other waste management strategies
- Use twentieth-century landfills to explain and illustrate the concepts of point-source pollution
- Use GIS technology to explore and analyze the human and environmental characteristics of an urban area
- Use GIS technology to map and analyze the effect of abandoned landfill sites on water resources and human activities in an urban area

GIS skills and tools

- Add data to the view
- Zoom to the active theme
- Change theme names
- Measure distance in a view
- Clear selected features
- Query the data

GEOGRAPHIC INQUIRY STEP	GIS SKILL
Ask a geographic question	• Recognize and understand geographic questions presented in a scenario.
Acquire geographic resources	• Add data to a new view to create a basemap of the study site that reflects its cultural and physical characteristics. • Add data reflecting the location and characteristics of abandoned landfill sites.
Explore geographic data	• Observe spatial patterns and relationships within and between basemap themes reflecting cultural and physical features. • Use the Legend Editor to symbolize census data to reveal patterns of population density. • Observe spatial relationships between landfill sites and the physical and cultural features of the study site using the Query Builder and select-by-theme functions. • Use the Query Builder and select-by-theme functions to analyze relationships between landfill sites and selected characteristics of the community.
Analyze geographic information	• Visually analyze spatial patterns of landfill sites in the community. • Create subsets of themes by using the selection features of the Query Builder (new set, select from set, add to set). • Use the Query Builder and select-by-theme functions to analyze relationships between landfill sites and selected characteristics of the community.
Act on geographic knowledge	• Create and print a layout to display the results of spatial analysis. • Prepare a report identifying hazardous landfill sites to present to municipal officials and environmental groups. • Make the presentation(s).

Teacher notes

Lesson introduction

Introduce this lesson by displaying a transparency of the graph called "Trends in waste generation, recovery, and disposal" (a color version of this graph is available in module 5 of *Community Geography: GIS in Action*). Give students about five minutes to write five observations about waste issues that are supported by the graph. Before beginning a discussion of student observations, spend a few minutes going over key terms: waste generation, waste recovery, waste disposal, recycling, combustion, composting, land disposal, point-source pollution, and nonpoint-source pollution. As students share their observations, write them in the space below the graph on the transparency. Observations should include the following:

- Between 1960 and 2000, the total amount of waste generated more than doubled from one hundred million tons annually to 220 million tons per year in the United States.
- Between 1960 and 2000, the amount of waste disposed of in landfills doubled from approximately sixty million tons per year to 120 million tons per year.
- Between 1960 and 2000, the relative percent of waste that goes to landfills or is burned dropped slightly while the percent of waste that is recycled or composted increased (composting does not appear as a significant disposal strategy until the mid-1980s).

To put the issue of landfills in a global perspective, be sure to spend some time discussing the following questions as well:

- Scientists use the phrase "waste sink" to describe the relationship between humans and the natural environment. What do you think that term means?

  The earth has to reabsorb the waste products of plant, water, animal, and human activity. The creation of large quantities of waste and nonbiodegradable products by modern human societies severely taxes the earth's ability to do this job.

- In what way is recycling different from the other three methods of waste disposal?

  All the other waste removal strategies require the earth to absorb human waste (the earth as a "waste sink")—only recycling does not.

- Of the disposal methods reflected in the graph, which ones are better for the earth? Which are worse?

  Better: Composting returns nutrients to the soil and recycling conserves the earth's resources.

  Worse: Landfills and combustion have the potential to introduce toxins into the earth's soil, water, and atmosphere.

- Which of these four methods of waste removal is used in our own community?

Tell the students that they are about to read a case study about the response of a group of Toronto high school students when they discovered that their city contained nearly seventy closed landfill sites. Instruct students to read the case study "Identifying potentially harmful landfills" in *Community Geography: GIS in Action*.

Discuss the following questions:

- What risks do abandoned landfill sites present to people?
- What risks do abandoned landfill sites present to the environment?
- Why are older landfills more dangerous than modern ones?
- What information would you want to have to evaluate the potential danger of a landfill site?

Exercise

Before completing this lesson with students, we recommend that you complete it as well. Doing so will allow you to modify the activity to accommodate the specific needs of your students.

Read the exercise scenario aloud with the class. Explain that in part 1 they will use GIS to create a basemap of the East York neighborhood and explore some of its physical and cultural characteristics. In part 2, they will add and explore data that locates landfill sites in the neighborhood. Finally, in part 3, they will identify the high-risk sites to be targeted first and prepare a map showing their location. Clarify any questions before the students begin to work individually.

TEACHER TIPS

✓ If your students are using ArcView 3.0, draw their attention to the directions that explain how to load appropriate legends for basemap themes.

✓ At the end of part 1, students will need to save their project. Be sure to give them instructions on how to rename the project and where they should save it in order to access it later.

✓ Advise students to follow the directions very carefully when working on steps that involve the Query Builder. It is essential to double-click the Fields and Values elements and single-click the Function element. If students are experiencing problems with their queries, it is usually because they did not click the elements correctly.

✓ Students should also follow the sequence of directions very carefully in the Select by Theme steps in parts 2 and 3.

✓ In part 3, students will be asked to create and save a shapefile of the high-risk landfill sites. As with saving the project, be sure to give them instructions on where they should save this new file.

✓ If time permits, encourage students to explore the data further.

  – Normalize the population in 1996 by percent of total and normalize the number of dwellings in 1996.

  – Determine how many landfill sites are located in park, industrial, or commercial areas.

  – Use a query in the theme properties to create separate landfill site themes for A3 and A5 sites.

Things to look for while the students are working on this activity:

• Do students understand the theme-on-theme selection concept? Unless they have a clear understanding of this function, they will not truly understand their results.

• Are the students experiencing any difficulty navigating between the table and view windows?

• Are the students answering the questions as they work through the procedure?

Conclusion

Conclude the lesson by discussing the seven high-risk sites identified in the exercise.

• Ask students to point out other landfill sites that seem particularly hazardous to them. Ask them to explain their choices.

• Ask students to suggest other possible definitions of high-risk sites that Toronto's Department of Environmental Management might have used.

Assessment

In the middle and high school assessments, students will apply what they learned in the case study and exercise by reconsidering criteria used to evaluate landfill risks. The assessment tasks will require them to demonstrate their knowledge of the potential risks posed by closed landfill sites and employ the power of GIS to analyze those risks in a specific study area.

Middle school: Highlights skills appropriate to grades 5 through 8

Middle school students will be asked to identify the medium- and lowest-risk sites that should be targeted in the second and third phases of the landfill sites study. They will be asked to do the following:

• Define and explain the criteria for site selection

• Identify and map sites that meet these criteria

• Predict how the landfills could affect the people of East York

High school: Highlights skills appropriate to grades 9 through 12

High school students will be asked to develop an alternate definition of high-risk sites as well as criteria to define medium- and lowest-risk sites in the neighborhood. They will be asked to do the following:

• Define and explain the alternate criteria for site selection

• Identify and map sites that meet these criteria

• Identify and explain a strategy for prioritizing the remaining sites

• Propose an action plan to raise public awareness of landfill hazards

TRENDS IN WASTE GENERATION, RECOVERY, AND DISPOSAL

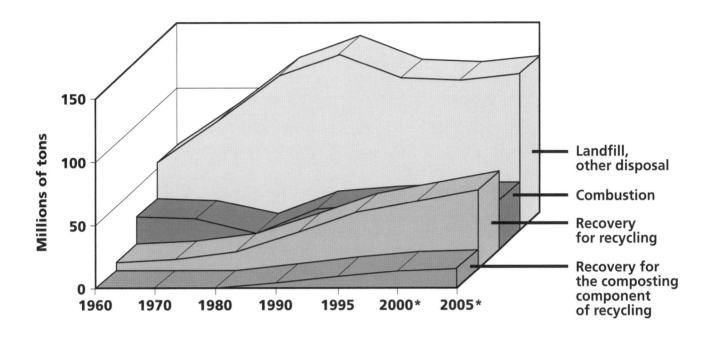

*Assumes 30% recovery in 2000 and 32% recovery in 2005.
Source: Characterization of MSW in the U.S.: 1998 Update, U.S. EPA, Washington, D.C.

STUDENT OBSERVATIONS:

MIDDLE SCHOOL ASSESSMENT

Map, query, and analyze neighborhood data to identify high-risk landfills

Name _____ Date _____

Toronto's Department of Environmental Management, impressed with your analysis of the sites posing the highest risk to the East York neighborhood, is now asking you to divide the remaining twenty-one sites into two further categories: medium risk and low risk. These sites will be targeted in the second and third phases of the department's landfill sites study. They will leave it up to you to determine the criteria for assigning sites to either category. When you have completed your analysis, prepare a report that includes:

1 Medium-risk and low-risk landfill criteria descriptions

- Criteria for how you defined medium-risk landfill sites
- Criteria for how you defined low-risk landfill sites
- A paragraph that explains your choice of criteria for medium- and low-risk sites

2 An essay that predicts how these landfills could affect the lives of people who live, work, and play in East York. Focus on issues you explored in the exercise such as proximity to schools and rivers.

For extra credit, create an action plan to raise public awareness of this issue.

3 A map that shows the location of the medium- and low-risk landfill sites

- It must include symbols with appropriate colors, sizes, shapes, and so forth that clearly communicate the message of the map.
- It must include a title, orientation, scale bar, author name, legend, and date.

MIDDLE SCHOOL ASSESSMENT RUBRIC

Map, query, and analyze neighborhood data to identify high-risk landfills

STANDARD	EXEMPLARY	MASTERY	INTRODUCTORY	DOES NOT MEET REQUIREMENTS
1 The student knows and understands how to make and use maps and databases to analyze spatial distributions and patterns.	Creates a map or series of maps that displays medium- and low-risk landfill sites. The map is symbolized appropriately and clearly communicates a message. It meets all listed requirements (e.g., appropriate themes, title, north arrow).	Creates an easy-to-read map that is symbolized appropriately, meets all listed requirements (e.g., appropriate themes, title, north arrow), and clearly identifies medium- and low-risk landfill sites.	Creates a map that identifies all medium- and low-risk landfill sites and meets most listed requirements (e.g., appropriate themes, title, north arrow). Attempts to symbolize the map appropriately.	Creates a map that identifies most but not all medium- and low-risk landfill sites, but is not symbolized appropriately or is missing several listed requirements (e.g., appropriate themes, title, north arrow).
4 The student knows and understands how to analyze the human characteristics of places.	Creates appropriate definitions of medium- and low-risk landfill sites based on the data provided in the exercise. The definitions are based on specific human characteristics of the neighborhood.	Creates reasonable definitions of medium- and low-risk landfill sties based on the data provided in the exercise. The definitions are linked to specific human characteristics of the neighborhood.	Attempts to create definitions for medium- and low-risk landfill sites, but the definitions are not closely related to the data from the exercise.	Attempts to create a definition of either medium- or low-risk landfill sites, but it does not relate to data from the exercise.
14 The student knows and understands the consequences of human modification of the physical environment.	Predicts how the various landfill sites will affect the people of East York and develops an action plan to raise public awareness of the issue.	Predicts how the various landfill sites will affect the people of East York.	Predicts how at least one type of landfill site will affect the people of East York.	Identifies why landfill sites are dangerous, but does not relate it to the scenario.
18 The student knows and understands how to apply the geographic point of view to solve social and environmental problems by making geographically informed decisions.	Presents a clear and reasonable explanation as to why their choice of criteria for medium- and low-risk landfill sites is logical based on the data.	Presents a clear explanation as to why their choice of criteria for medium- and low-risk landfill sites is logical.	Attempts to present an explanation of their choice of criteria for medium- and low-risk landfill sites, but it is not based on the data.	Presents little or no explanation of their choice of criteria for medium- and low-risk landfill sites.

This is a four-point rubric based on the National Standards for Geographic Education. The "Mastery" level meets the target objective for grades 5–8.

HIGH SCHOOL ASSESSMENT

Map, query, and analyze neighborhood data to identify high-risk landfills

Name _____   Date _____

Toronto's Department of Environmental Management has been told that they must reconsider the criteria they used to identify landfill sites that pose the highest risk to East York. Critics believe that their initial criteria overlooked a number of landfill sites that should also be designated high-risk. Your job is to develop new criteria to identify high-risk, medium-risk, and lowest-risk sites. When you have completed this task, prepare a report that includes:

1 High-risk, medium-risk, and low-risk landfill criteria descriptions
  - A description of how you redefined the high-risk landfill sites
  - A description of how you defined medium-risk landfill sites
  - A description of how you defined low-risk landfill sites
  - An explanation of your choice of criteria for all risk levels of landfill sites

2 Action plan to raise public awareness of the high-to-medium-risk landfill sites, and list of changes that need to take place to keep people safe from possible hazards
  - This plan could include letters to the residents, advertisements on local radio and television stations, neighborhood meetings, and so on.

3 A map that shows the location of high-, medium-, and low-risk landfill sites
  - It must include symbols with appropriate colors, sizes, shapes, and so forth that clearly communicate the message of the map.
  - It must include a title, orientation, scale bar, author name, legend, and date.

HIGH SCHOOL ASSESSMENT RUBRIC

Map, query, and analyze neighborhood data to identify high-risk landfills

STANDARD	EXEMPLARY	MASTERY	INTRODUCTORY	DOES NOT MEET REQUIREMENTS
1 The student knows and understands how to use geographic representations and tools to analyze, explain, and solve geographic problems.	Creates a map or series of maps that displays appropriately defined high-, medium-, and low-risk landfill sites. The map is symbolized appropriately and clearly communicates a message. It meets all listed requirements (e.g., appropriate themes, title, north arrow).	Creates an easy-to-read map that is symbolized appropriately, meets all listed requirements (e.g., appropriate themes, title, north arrow), and identifies reasonable high-, medium-, and low-risk landfill sites.	Creates a map that identifies some high-, medium-, and low-risk landfill sites and meets some listed requirements (e.g., appropriate themes, title, north arrow). Attempts to symbolize the map appropriately.	Creates a map that identifies few high-, medium-, and low-risk landfill sites and is missing several listed requirements (e.g., appropriate themes, title, north arrow) or is not symbolized appropriately.
4 The student knows and understands the changing physical and human characteristics of places.	Creates an innovative campaign to inform the public about landfill hazards, and provides a list of changes that need to be made to make the public safe. The action plan includes a variety of methods to raise public awareness.	Proposes changes to help protect the public from, and inform it about, hazards associated with landfill sites. The action plan includes at least two methods to raise public awareness.	Proposes changes to help protect the public from hazards associated with high-risk landfill sites, but does not list changes that need to be made to keep the public safe.	Lists some dangers of landfill sites, but does not propose any changes or suggest methods to raise public awareness.
14 The student knows and understands how to apply appropriate models and information to understand environmental problems.	Redefines high-, medium-, and low-risk criteria based on GIS data and outside research.	Uses GIS data to redefine high-, medium-, and low-risk landfill sites.	Uses GIS data and projects to redefine medium- and low-risk landfill sites.	Defines only one of the three types of landfill risk and does not base the definition on the GIS data.
18 The student knows and understands how to use geographic knowledge, skills, and perspectives to analyze problems and make decisions.	Presents a clear and reasonable explanation as to why their choice of criteria for high-, medium-, and low-risk landfill sites is the most logical based on the data. Provides evidence that the public awareness campaign is appropriate.	Presents a clear explanation as to why their choice of criteria for high-, medium-, and low-risk landfill sites is logical. Explains why the public awareness campaign is appropriate.	Attempts to present an explanation of their choice of criteria for high-, medium-, and low-risk landfill sites, but it is not based on the data. Attempts to explain why the public awareness campaign is appropriate.	Presents little or no explanation of their choice of criteria for high-, medium-, and low-risk landfill sites. Provides little or no explanation for why the public awareness campaign is appropriate.

This is a four-point rubric based on the National Standards for Geographic Education. The "Mastery" level meets the target objective for grades 9–12.

On your own

Overview

This section provides guidelines and information to help you implement a similar project in your own classroom. Point-source pollution is a potential threat to communities throughout the world. Analyzing pollution point sites such as abandoned landfills is an excellent way to teach students key problem-solving skills: how to locate appropriate data with which to study the problem, how to integrate information from various sources, how to analyze patterns and relationships within data, how to collaborate with others in decision making, and how to communicate one's findings in a variety of formats. With access to data and the power of GIS technology, students—the decision makers of the future—have the power to shape decisions made by others today.

Estimated time

The time it will take to study a point-source pollution issue in your own community will vary with the complexity of the issue you study. The landfill study undertaken by the students at Toronto's Crescent School took three weeks to complete. Their final presentation, however, took place several months later.

Materials

- Computers for all students (one per student or student group)
- Data reflecting physical and human characteristics of the study site
- Data reflecting the location and characteristics of possible pollution points
- Access to a color printer
- Class e-mail list for correspondence with community partners
- Reserved disk space for data storage

Standards

GEOGRAPHY STANDARD	MIDDLE SCHOOL	HIGH SCHOOL
1 How to use maps and other geographic representations, tools, and technologies to acquire, process, and report information from a spatial perspective	The student knows and understands how to make and use maps and databases to analyze spatial distributions and patterns.	The student knows and understands how to use geographic representations and tools to analyze, explain, and solve geographic problems.
3 How to analyze the spatial organization of people, places, and environments on Earth's surface	The student knows and understands distributions of physical and human phenomena with respect to spatial patterns, arrangements, and associations.	The student knows and understands how to analyze relationships in and between places.
4 The physical and human characteristics of places	The student knows and understands how physical and human processes together shape places.	The student knows and understands the changing physical and human characteristics of places.
12 The processes, patterns, and functions of human settlement	The student knows and understands the internal spatial structure of urban settlements.	The student knows and understands the functions, sizes, and spatial arrangements of urban areas.
14 How human actions modify the physical environment	The student knows and understands that the physical environment can both accommodate and be endangered by human activities.	The student knows and understands the global effects of human changes in the physical environment.
17 How to apply geography to interpret the past	The student knows and understands how the spatial organization of a society changes over time.	The student knows and understands how the processes of spatial change affect events and conditions.
18 How to apply geography to interpret the present and plan for the future	The student knows and understands how to apply the geographic point of view to solve social and environmental problems by making geographically informed decisions.	The student knows and understands how to use geographic knowledge, skills, and perspectives to analyze problems and make decisions.

Objectives

The student is able to:

- Acquire and prepare data for analysis in a GIS
- Use GIS to identify the human and physical features of a community
- Identify and analyze the human and physical features of a community that are at risk from nearby sources of pollution

GIS skills and tools

GEOGRAPHIC INQUIRY STEP	GIS SKILL
Ask a geographic question	• What are the potential causes of point-source pollution in your own community? • Which neighborhoods and natural resources are at greatest risk from point-source pollution in your community?
Acquire geographic resources	• Work with local, county, and state GIS users to obtain physical and cultural data for your proposed study site. • Create a basemap of the study site with data that reflects its cultural and physical characteristics. • Work with environmental groups or government agencies to obtain data about point-source pollution sites relevant to your investigation.
Explore geographic data	• If necessary, geocode point-source pollution sites to add them to an ArcView project or add them to the project as an event theme. • Symbolize basemap data in a variety of ways to explore characteristics of the study site. • Observe spatial relationships between point-source pollution sites and the physical and cultural features of the study site using queries and spatial selection.
Analyze geographic information	• Visually analyze spatial patterns of point-source pollution sites in the community. • Create subsets of themes by using the selection features of the Query Builder (new set, select from set, add to set). • Make copies of relevant themes and use the Theme Definition function in Theme Properties to display a specific characteristic of that theme. • Use theme-on-theme selection to analyze relationships between point-source pollution sites and selected characteristics of another theme.
Act on geographic knowledge	• Create and print a layout to display the results of spatial analysis. • Prepare a report identifying hazardous point-source pollution sites to present to municipal officials and environmental groups. • Educate others about your results.

Teacher notes

Ask a geographic question

The Crescent School investigation of landfills originated with student interest in a local situation that was widely reported in the news—the contamination of the community of Walkerton's well water from a nearby livestock operation. The students' interest in this story led them to the discovery that abandoned landfills in their own neighborhood could potentially contaminate their own city's water resources. In creating your own community study, a variety of events can foster the development of a central question for your students. For example:

- An article in local newspapers or other media about existing or potential danger from a point-source pollution site or sites. (Even if your own community does not face an identical danger, the article can serve as a springboard to greater awareness of similar dangers close to home just as the Walkerton story did for Crescent School students.)
- A specific event (an oil spill, for example) that raises community awareness of a potential threat from a point-source pollution site or sites.

One way to develop an appropriate question for the focus of your own community study is to begin with student research into local environmental issues and concerns:

- Assign students to monitor newspapers and television news programs for two weeks (or a longer period of time) to identify environmental issues of local concern.
- Assign students to review newspapers from the previous year (or a different period of time) to identify environmental issues of local concern.
- Have students conduct a survey in the community to identify environmental issues of local concern.
- Assign students to research the environmental impact of local businesses like gas stations, dry cleaners, golf courses, and factories on the surrounding environment.

As a teacher, you can also direct student awareness toward a particular issue rather than waiting for the "teachable moment" to arise on its own:

- Invite a guest speaker from, or take a field trip to, a local environmental organization to raise student awareness of the potential threat from a point-source pollution site.
- Foster student awareness by providing appropriate reading materials, films or videos, and Internet resources related to point-source pollution.
- Raise student awareness by creating a point-source pollution "scavenger hunt" to have students locate important facts from a variety of sources.
- Assign relevant research topics in point-source pollution to your students (or to groups of students) and have them prepare posters, brochures, or multimedia presentations to share their findings.

Whatever the process that raises student awareness of or triggers their interest in point-source pollution, the intent of that process is to motivate them to become personally involved in confronting this issue in their own community. Build your study around an essential question such as "Which rivers and neighborhoods are at greatest risk from abandoned landfills in East York?" In the process of research and exploration, that question can be broken down into a series of more limited preliminary questions (see *Community Geography: GIS in Action,* page 143) that will guide you in your analysis.

Acquire geographic resources

TEACHER TIPS ON OBTAINING DATA ABOUT YOUR COMMUNITY

✓ See *Community Geography: GIS in Action* (page 144) for a discussion of possible sources of community data for your study.

✓ Be sure to explain to potential providers that the data will be used for a school project. In some cases data providers will waive data fees for an educational project or allow use of data that is not otherwise distributed. Be sure to acknowledge all data providers in any presentation about your project.

✓ If you are able to limit the size of your study area it may be easier to acquire the necessary data. Many data providers who are reluctant to release large data sets will willingly clip and grant permission to use a subset of that data.

TEACHER TIPS ON OBTAINING DATA ABOUT POINT-SOURCE POLLUTION SITES

✓ See *Community Geography: GIS in Action* (page 145) for a discussion of possible sources of point-source pollution data for your study.

✓ Environmental groups often lack the technology or the people power to carry out investigative studies of their own. The resources that you bring to the investigation of a specific pollution issue (computers and students) can form the basis for a very powerful partnership with such groups. Do not hesitate to reach out to local groups.

Explore geographic data

Technical issues

A number of variables will influence how you go about setting up the GIS basemap for your study of a local point-source pollution issue: the number of students in your class, the number of computers at your disposal, and the amount of time you can devote to the project. Depending on your situation, there are many possibilities.

Ideas for setting up your basemap:

- Provide all students with a completed basemap that you, a student assistant, or a small group of students has prepared in advance.
- Have each student or group create their own basemap with data that you provide. Establish conventions for cartographic issues such as labeling and symbolization.
- Let students or groups create their own basemap using preestablished legends for themes such as land use, roads, and landmarks so that maps prepared by different students or groups are easily compared.

The point-source pollution site data must be in shapefile format in order to display it on your basemap in ArcView. If the site data that you acquire is not already in shapefile format, you will need to prepare a data table and add it to your project. This can be accomplished by:

- Geocoding if you have street addresses for your sites (see module 2, "Reducing crime")
- Adding an event theme if you have latitude and longitude coordinates for each site (see module 4, "Tracking water quality")

Ideas for preparing point-source pollution site data:

* Students or groups create their own basemap for the project and then create a shapefile for point-source pollution sites by geocoding or adding an event theme.
* Provide all students with a point-source pollution sites shapefile that either you or another student has already created.

Preliminary data exploration

* To guide students in their preliminary exploration of project data, ask them to observe and record spatial patterns in each of the data sets (human characteristics of the community, physical characteristics of the community, point-source pollution sites).
* Encourage students to use the attribute tables in their exploration. Sorting the tables according to different data fields and then selecting particular subsets of data will often reveal spatial patterns that were not otherwise apparent.
* Encourage students to look for relationships between data sets such as proximity of features to one another or population patterns in different land-use categories.
* Use a projection device to demonstrate some of the techniques students can use to explore the data.
  - Sort and select features from the attribute table.
  - Use the theme definition function to create new themes displaying a subset of theme data.
  - Use graduated color to symbolize various population characteristics.
  - Use the Select by Theme function to explore the proximity of basemap features to point-source pollution sites and vice versa.
  - Perform queries.
* Discuss and compare spatial patterns that students have observed before beginning the actual analysis. This will allow students to learn from each other and foster ideas for further exploration. Discuss whether the conclusions are valid and supported by the data.

Analyze geographic information

The essential question of your investigation and its underlying preliminary questions will shape the analysis of your study site and the articulation of your findings. The final analysis will also be guided by conclusions reached during the preliminary exploration of the data. Typically, a study of this kind results in a report or presentation of findings to an appropriate audience. Student analysis, then, should be presented in a report with conclusions that are expressed clearly and are supported by the data. There is a range of possibilities for the analysis phase of your study.

* Student grouping
  - Students may work individually, in pairs, or in small groups for the analysis of information.
  - If your class is small, they may work together as one team and divide the analysis tasks within the group.
  - Different students or groups may work on separate aspects of the investigation. For example, one group may focus on the relationship between the point-source pollution sites and natural resources while another group focuses on the population patterns near the point-source pollution sites.
* Provide student guidelines
  - Specify the questions that students should answer in their analysis.
  - Specify categories of information that students should cover in their analysis (population characteristics, classification of point-source pollution sites, and so on).
  - Specify materials (e.g., maps, data tables, charts) students should prepare to support their conclusions.
  - If the analysis will lead to recommendations, give students specific instructions for these as well.
  - Provide a rubric for the completed report or presentation so that students know exactly what is expected of them.

Act on geographic knowledge

The final step in the geographic inquiry process is the one too often overlooked. It is only by acting on their findings that students will experience the process and pride of responsible citizenship. Through personal experience, they will understand that one small group of people really can make a difference and that they can make a meaningful contribution to their community now.

- Prepare a multimedia presentation of your findings and recommendations and ask if you and your students can give the presentation to such audiences as:
  - Meetings of local or regional environmental groups
  - Meetings of municipal or county government bodies
  - Meetings of local service organizations (Rotary, Kiwanis, and so on)
- Invite appropriate officials and special-interest groups to your school for a presentation of your findings and recommendations. Write a press release to inform local media about the presentation.
  - Local, county, or state environmental agencies
  - Local, county, or state legislators
  - Local, county, or state environmental protection groups
- Prepare a presentation and lesson for younger children in your community and ask if you can visit their classroom to present it.
- Prepare a brochure or poster to illustrate your conclusions and recommendations and distribute it locally.

Lesson

Case study: Who walks and who takes the bus?
Exercise: Use buffers to identify eligible school-bus riders

Overview

This lesson focuses on issues of school transportation planning and the spatial relationships among geographic features. The exercise in this lesson has two parts. In part 1, students will draw circular buffer zones according to the school's bus transportation guidelines. They will select student point locations using a combination of spatial and attribute queries. In part 2, students will repeat the analysis, this time using the network-based zones from the case-study project. Finally, students will complete a table showing the number of eligible bus riders by grade level, estimate the number of buses needed, and create a layout showing the locations of eligible bus riders. In the assessment, students will use the skills they learned to analyze new guidelines (middle school) or to develop new guidelines (high school).

Estimated time

45-MINUTE CLASS	ACTIVITY
1	Exercise introduction
	Exercise part 1
2	Exercise part 2
3	Assessment

Materials

- *Community Geography: GIS in Action* book (one per student)
- Computers for all students (for the exercise and assessment)
- Color printer (to print the student maps)

Student handouts from this lesson to be photocopied:

- Assessment

Standards

GEOGRAPHY STANDARD	MIDDLE SCHOOL	HIGH SCHOOL
1 How to use maps and other geographic representations, tools, and technologies to acquire, process, and report information from a spatial perspective	The student knows and understands how to make and use maps, globes, graphs, charts, models, and databases to analyze spatial distributions and patterns.	The student knows and understands how to use geographic representations and tools to analyze, explain, and solve geographic problems.
3 How to analyze the spatial organization of people, places, and environments on Earth's surface	The student knows and understands how to use the elements of space to describe spatial patterns.	The student knows and understands the spatial behavior of people.
5 That people create regions to interpret Earth's complexity	The student knows and understands the elements and types of regions.	The student knows and understands how to use regions to analyze geographic issues.
18 How to apply geography to interpret the present and plan for the future	The student knows and understands how to apply the geographic point of view to solve social problems by making geographically informed decisions.	The student knows and understands how to use geographic knowledge, skills, and perspectives to analyze problems and make decisions.

Objectives

The student is able to:

- Create buffer zones around a school according to distance
- Map and analyze student data according to distance buffer zones and grade level
- Determine which students are eligible to ride the bus based on the school's transportation guidelines

GIS skills and tools

☑	Zoom to the active theme	✋	Pan the view
⊙	Draw a circle to create buffers	▦	Select all records in a table
▣	Clear features that are selected	▦	Calculate attribute values
▦	Open a theme's attribute table	◹	Select features using a query expression
⟳	Switch selected records in a table	Σ	Summarize attribute field
✚	Add a theme to the view		

GEOGRAPHIC INQUIRY STEP	GIS SKILL
Ask a geographic question	• Develop geographic questions from a scenario containing geographic criteria.
Acquire geographic resources	• Get these from the project data directory. • Create transportation buffer zones according to the school's guidelines provided in scenario. • Add data to a project. • Subtract features of a polygon theme.
Explore geographic data	• Observe the spatial relationships between themes. • Explore attributes using an attribute table.
Analyze geographic information	• Add a new field and calculate value. • Query the student data by grade. • Use theme-on-theme selection.
Act on geographic knowledge	• Determine the number of students by grade level who are eligible to ride the bus. • Create a layout to illustrate the location of eligible students by zone.

Teacher notes

Lesson introduction

Begin the lesson by having the students read the case study "Who walks and who takes the bus?" Discuss the case study with the class. During the discussion, point out to students that regardless of whether a student rides a bus, takes a car, or walks, there is planning involved in deciding when to leave home and what route to take to get to school. Because school buses must pick up multiple students from many locations, extensive planning is necessary to make the routes functional and efficient.

To help students understand the power of GIS in solving this seemingly simple problem, have them imagine the following scenario:

1 They are in charge of bus transportation for the school.

2 They have a list of all students, their addresses, and their ages.

3 They must determine from this list who is eligible to ride the bus and who is not, based on the following factors:

- Those who live within five blocks of the school cannot ride
- Those who can drive (age sixteen or older) may choose not to ride
- Those who live farther than five blocks and are under sixteen are eligible to ride

How could they go about solving this problem? Using pen and paper to solve the problem could be a daunting task. A list of addresses, names, and ages does not provide you adequate information to find a solution. To solve the problem you must have a map and the data to analyze together. The most efficient means to do this? GIS!

Next, have students complete the exercise. Explain that they will follow the steps that the high school students in Ottawa did to assist the elementary school principal in determining who is eligible to ride the bus.

Exercise

Before completing this lesson with students, we recommend that you complete it as well. Doing so will allow you to modify the activity to accommodate the specific needs of your students.

Read the exercise scenario aloud with the class. Explain that the goal of the exercise is to help the elementary school principal determine which students are eligible to ride the bus. To do this they must first create transportation zones based on the school's busing guidelines. They will then compare student data against these guidelines and zones to identify and locate students who are eligible to ride the bus. Ultimately, they will provide the principal with a list of eligible students by grade level and a map showing the locations where the students live. Clarify any questions before the students begin to work individually.

TEACHER TIPS

✓ You should have the students save their projects regularly. More information about saving and renaming projects can be found in the module 1 "Teacher tips" section.

✓ In part 1, steps 6 and 10, students will need to save the shapefiles they create. Be sure to give them instructions as to where they should save these new files. The software's default is to save new files in the Windows "Temp" folder on the C: drive. If students save their files here, they won't be able to find them again if they are working on a different computer.

✓ In part 1, questions 16a and 19a, students will get slightly different answers depending on the exact size and center point of their buffers. They will be able to measure the radius more precisely if they enlarge their view window (thus enlarging the map scale without zooming in).

✓ Advise students to follow the directions very carefully when working on steps that involve the Query Builder. It is essential to double-click the Fields and Values elements, and single-click the Function element. If students are experiencing problems with their queries, it is usually because they didn't click the elements correctly. Be sure that the students are checking their expression with the graphic in the exercise instructions.

Things to look for while students are working on this activity:

- Are the students saving their work periodically to the designated folder?
- Are the students able to move between tables and views and resize windows appropriately?
- Are the students developing additional geographic questions based on their findings?
- In part 2, steps 9–13, are the students successful in adding a new field and populating it using the Calculate tool?
- In part 2, step 15, are the students successful in using the Query Builder?
- In part 2, step 20a, did students come up with the correct Boolean operators for the query expression? It is important for students to have the correct expression because they will be using it in step 21.

Conclusion

Once students have finished the exercise, have them save the project so that they may refer to it for the assessment. Because everyone is working with the same data and guidelines, they should have all come to similar conclusions at the end of the exercise. Discuss the answers to the questions presented in the exercise (refer to the answer key at the back of this book). Share the final layout as a group to make sure all students successfully completed the analysis. Possible discussion items include:

- Have students explain the logic used in the exercise to create the various selection sets.
- What alternative methodologies could be used to produce the same results?
- What patterns do they see, if any, in the distribution of students who are (or are not) eligible to ride the bus? How might they approach the problem of assigning groups of students to individual buses?
- If the school guidelines were changed, how would they update their analysis?

Assessment

Assign the students to work in teams of three for the assessment. Students will create a map and a formal presentation to the class as though their classmates were the school transportation committee. The map will be created from the exercise data and analysis, so students will need access to the computers and, optionally, to a printer. In the final presentation it would be best if students exported the views into a multimedia presentation tool so that the entire class can easily view them. One way to conclude the presentations is to have students vote on the best transportation plan. See "On your own: Project planning" in *Community Geography: GIS in Action* for additional strategies for presentations.

There are a number of alternate formats for submitting the final student assessments:

- Multimedia presentation
- Illustrated brochure
- Illustrated text (word-processor) document

Middle school: Highlights skills appropriate to grades 5 through 8

Each small group must work together as a team to develop a bus transportation plan using new busing guidelines. They will analyze the student data with the new buffers and then present the revised map and data to the class, including summary statistics on who is eligible to ride the bus.

High school: Highlights skills appropriate to grades 9 through 12

The group is presented with the challenge of developing new busing guidelines so that an additional fifty to sixty students walk to school. This will eliminate the cost of one bus from the school's budget and allow more students to get daily exercise by walking to school. Each small group must develop and test their new guidelines using the student data and project from the exercise. Each group will present their solution to the class and include a map and table showing that the number of bus riders has decreased compared to the original guidelines.

MIDDLE SCHOOL ASSESSMENT

Use buffers to identify eligible school-bus riders

Name _____ Date _____

The principal of the elementary school needs your help once again. The school board has changed the transportation guidelines for busing students. The principal has asked you to create a map showing where the students live who are eligible to be bused according to the new guidelines. Here are the new guidelines:

Busing guidelines

GRADE	DISTANCE WHEN STUDENTS ARE BUSED
Junior and senior kindergarten (JK and SK)	All bused regardless of distance from school
Grades 1–3	Live 1,500 meters or more from school (previously was 1,000 meters or more)
Grades 4–6	Live 2,000 meters or more from school (previously was 1,600 meters or more)

Part 1. Create a transportation map

1 Use the exercise project and student data as your basemap.

2 Use the exercise as a guide:
- Are your geographic questions the same?
- Create distance buffers using the draw polygon tool.
- Analyze student data using the following tools and your new distance buffers:
  - Select by Theme
  - Add Field
  - Query Builder
  - Calculate

3 Create a layout of your map and include the following information:
- 1,500- and 2,000-meter buffers
- Locations of the school and student residences
- Table of summary statistics
- Orientation (compass rose)
- Author(s)
- Date

Part 2. Present your findings

You will present your map to the school transportation committee (your class). You will explain your map and how you determined your findings. You should include the following items in your presentation:
- A map showing the location of the new buffers and where bus riders live
- Table indicating the number of students eligible to ride the bus and the total number of students
- Explanation of your map
- Explanation of how you determined your findings
- Comparison of the number of students bused using the old and new guidelines

MIDDLE SCHOOL ASSESSMENT RUBRIC

Use buffers to identify eligible school-bus riders

STANDARD	EXEMPLARY	MASTERY	INTRODUCTORY	DOES NOT MEET REQUIREMENTS
1 The student knows and understands how to make and use maps, globes, graphs, charts, models, and databases to analyze spatial distributions and patterns.	Creates a map that accurately represents the new guidelines and meets all listed requirements (busing zones, student locations, title, north arrow, and so on). Creates a table that lists the number of eligible students in each grade level for the new and old guidelines.	Creates an easy-to-read map that accurately represents the new guidelines and meets all listed requirements (busing zones, student locations, title, north arrow, and so on). Creates a table that lists the number of eligible students in each grade.	Creates a map that identifies the new busing zones and meets most listed requirements (busing zones, student locations themes, title, north arrow, and so on).	Creates a map that includes an inaccurate representation of the new busing zones and is missing several listed requirements (student locations, title, north arrow, and so on).
3 The student knows and understands how to use the elements of space to describe spatial patterns.	Analyzes student data to determine which students will ride the bus based on their grade and distance from school. Is able to describe the relationships of the different zones to the school and student locations on the map.	Analyzes student data to determine which students will ride the bus based on their grade and distance from school.	Analyzes student data based on the new guidelines but determines the location and number of bus riders in only one zone or grade level.	Creates new zones but cannot determine which students are eligible to ride the bus.
5 The student knows and understands the elements and types of regions.	Creates regions on a map accurately illustrating the new busing guidelines based on distance from school. Compares and contrasts new and old regions both visually and orally.	Creates regions on a map accurately illustrating the new busing guidelines based on distance from school.	Creates regions on a map according to the new guidelines, but they are somewhat inaccurate.	Creates at least one of the new regions, or has highly inaccurate regions that do not coincide with the new guidelines.
18 The student knows and understands how to apply the geographic point of view to solve social problems by making geographically informed decisions.	Correctly determines the number of students eligible to be bused under the new guidelines. Compares new and old guideline results and explains how they determined their findings.	Correctly determines the number of students eligible to be bused under the new guidelines. Compares new and old guideline results.	Determines the number of students eligible to be bused under the new guidelines, but they may be incorrect. Gives a general comparison of new and old guideline results.	Attempts to determine the number of eligible students but does not complete the analysis or has incorrect results. Does not compare new and old guideline results.

This is a four-point rubric based on the National Standards for Geographic Education. The "Mastery" level meets the target objective for grades 5–8.

HIGH SCHOOL ASSESSMENT

Use buffers to identify eligible school-bus riders

Name _____   Date _____

The principal of the elementary school needs your help once again. The school board has determined that the school needs to eliminate one school bus because of budget constraints, and they also want to have more students getting daily exercise by walking to school. The school transportation committee must create new guidelines so that fifty to sixty students who did ride the bus now walk. The principal has asked you and your team to use the data from your earlier project to suggest new busing guidelines. You will need to produce a map illustrating the new guidelines and present your findings to the school's transportation committee.

Part 1. Create a transportation map

A Use the exercise project as your basemap.

B Experiment with different distance buffers using the draw polygon tool as you did in part 1 of the exercise (refer to the exercise if you need help).

C Determine new distance rules based on your experimentation.

D Test your new guidelines:
- Create distance buffers based on the new guidelines.
- Analyze the student data using the new buffers.
- Compare your findings with the original data from the exercise to see if you reduced the number of bus riders by fifty to sixty students.

E Create a layout of your map and be sure to include the following information:
- Appropriate themes such as new buffers and student locations
- Table comparing old and new guideline results
- Title
- Orientation (compass rose)
- Author(s)
- Date

Part 2. Present your findings

You will present your map to the school transportation committee (your class). You will explain your map and how you determined your findings. Address the following items in your presentation:
- How did you develop the new guidelines? Describe the process step by step.
- What makes the new guidelines successful?
- Why should your solution be chosen?
- Illustrate your solution in map and table format.
- Compare the old and new guidelines in map and table format.

HIGH SCHOOL ASSESSMENT RUBRIC

Use buffers to identify eligible school-bus riders

STANDARD	EXEMPLARY	MASTERY	INTRODUCTORY	DOES NOT MEET REQUIREMENTS
1 The student knows and understands how to use geographic representations and tools to analyze, explain, and solve geographic problems.	Creates a map that accurately represents the new guidelines and meets all listed requirements (busing zones, student locations, title, north arrow, and so on). Creates a table that lists the number of eligible students for the old and new guidelines.	Creates an easy-to-read map that accurately represents the new guidelines and meets all listed requirements (busing zones, student locations, title, north arrow, and so on). Creates a table that lists the number of eligible students in each grade.	Creates a map that identifies the new busing zones and meets most listed requirements (busing zones, student locations themes, title, north arrow, and so on).	Creates a map that includes an inaccurate representation of the new busing zones and is missing several listed requirements (busing zones, student locations, title, north arrow, and so on).
3 The student knows and understands the spatial behavior of people.	Analyzes student data to develop new guidelines so that fifty to sixty fewer students are eligible to ride the bus. Develops new guidelines that are the most appropriate for each student age group.	Analyzes student data to develop new guidelines so that fifty to sixty fewer students are eligible to ride the bus. Develops new guidelines that are reasonable for each student age group.	Analyzes student data to develop new guidelines, but is only able to have thirty to forty fewer students riding the bus or uses unreasonable criteria.	Analyzes student data to develop new guidelines, but does not meet the requirements of fifty to sixty fewer students riding the bus.
5 The student knows and understands how to use regions to analyze geographic issues.	Creates regions on a map accurately illustrating the new busing guidelines based on distance from school. Gives a detailed comparison, both visually and orally, of the old and new guideline results.	Creates regions on a map accurately illustrating the new busing guidelines developed in the new plan. Performs a basic comparison of the results of the old and new guidelines.	Creates new regions on a map, but they do not accurately reflect the new guidelines developed. Attempts to compare the results of the old and new guidelines.	Has difficulty creating new regions that accurately reflect either the old or new guidelines. Does not compare the results of the old and new guidelines.
18 The student knows and understands how to use geographic knowledge, skills, and perspectives to analyze problems and make decisions.	Presents a clear argument as to why their guidelines are the most reasonable. Is able to articulate the geographic inquiry process they went through to develop the solution. Is able to describe the GIS skills and tools used to analyze the data.	Presents a clear argument as to why their guidelines are reasonable. Is able to describe the process they used to analyze the data and develop the solution.	Presents a solution to the problem, providing some evidence, but does not articulate the use of geographic skills or tools.	Presents a solution, but does not provide evidence for the solution.

This is a four-point rubric based on the National Standards for Geographic Education. The "Mastery" level meets the target objective for grades 9–12.

On your own

Overview

This section provides guidelines and information to help you implement a school-bus routing project with your students. This type of project allows students to give back to the educational community of which they are a part. Not only will students gain an appreciation for the challenges of solving transportation problems with GIS, they will have fun using their knowledge and problem-solving skills to positively affect a local school—perhaps their own.

Estimated time

In the case study, the students in the geomatics class dedicated two months of work to their project, spending about eighty minutes per class. This included the entire geographic inquiry process from developing the questions, meeting with partners, obtaining the data, and analyzing the information, to creating a final product for presentation.

When crafting your geographic question, take into consideration the time you have available for your project. For example, if you have one week for the project, you could limit the project to a basic geocoding activity or to determining general busing guidelines based on locations that have already been geocoded. An important factor to keep in mind is the length of time needed to obtain permission to use student data—this took three weeks in the case-study project.

Materials

- Computers and appropriate GIS software for all students
- Access to a color printer or plotter to print out maps (optional)
- Student data
- Geocodable, current street data
- Class e-mail address for communication with community partners
- Reserved disk space for data storage

Standards

GEOGRAPHY STANDARD	MIDDLE SCHOOL	HIGH SCHOOL
1 How to use maps and other geographic representations, tools, and technologies to acquire, process, and report information from a spatial perspective	The student knows and understands how to make and use maps, globes, graphs, charts, models, and databases to analyze spatial distributions and patterns.	The student knows and understands how to use geographic representations and tools to analyze, explain, and solve geographic problems.
3 How to analyze the spatial organization of people, places, and environments on Earth's surface	The student knows and understands how to use the elements of space to describe spatial patterns.	The student knows and understands the spatial behavior of people.
5 That people create regions to interpret Earth's complexity	The student knows and understands the elements and types of regions.	The student knows and understands how to use regions to analyze geographic issues.
18 How to apply geography to interpret the present and plan for the future	The student knows and understands how to apply the geographic point of view to solve social problems by making geographically informed decisions.	The student knows and understands how to use geographic knowledge, skills, and perspectives to analyze problems and make decisions.

Objectives

The student is able to:

- Create a map from tabular data
- Create buffer zones defined by distance
- Create maps illustrating the spatial interpretation of the busing guidelines (distance zones)
- Analyze the student locations in relation to the zones
- Identify and create maps of students who are eligible to ride the bus and those who must walk

GIS skills and tools

GEOGRAPHIC INQUIRY STEP	GIS SKILL
Ask a geographic question	• Develop one or more geographic questions related to transporting students to school or another transportation service in your community.
Acquire geographic resources	• List data needed to answer the geographic questions. • List any criteria needed for analysis (e.g., a school's busing guidelines). • Identify and obtain data from reliable sources. • Add student data to the GIS project. • Reproject map data if necessary.

Explore geographic data	• Geocode student data.
	  – Make a theme matchable.
	  – Locate an address.
	  – Match a list of addresses.
	• Correct any unmatched addresses.
	• Observe the spatial relationships between themes.
	• Explore attributes using the Identify tool and attribute table.

Analyze geographic information	• Create buffers or service areas. • Select students by attribute query (e.g., by grade) and spatial query (e.g., by zone). • Add new field (e.g., Bus) and calculate attributes (e.g., yes or no) based on criteria.
Act on geographic knowledge	• Create layouts and presentations that communicate findings or recommend actions. • Prepare and deliver data (reports, data, or interactive GIS maps) usable by your partner (e.g., school administrators).

Teacher notes

Ask a geographic question

TEACHER TIPS

✓ Have students meet with the school principal and transportation director who you will be partnering with on the transportation project. When they meet, have the students explain—and if possible, demonstrate—what GIS can do to assist them with their transportation needs. For example, you could demonstrate some of the exercise steps to show how you could analyze the data. Often, as in the case study, this will spark further ideas and the principal or other partner will be able to help the students better form their geographic questions.

✓ Clearly identify what the transportation guidelines are and analyze each piece. This will help students better understand their geographic question. It will also help them identify what resources they need. Refer to the module 6 case-study section "Identifying the problem" (page 152) for details on how this was done in the case study.

✓ Have students discuss alternative scenarios. For example, what factors would be different in creating buffer zones and planning busing for a middle school rather than an elementary school? What about for a high school or a community college? What about for a magnet school, one whose students seldom live in the surrounding neighborhood, or an urban school where students take public transportation instead of school buses?

✓ Consider the number of students in your class and how you would like them to work. For example, if you have them work in teams, do you want each team to explore a different question and then work together as a class to construct a unified plan? Or do you want each team to try to solve the same problem but come up with a variety of solutions?

Acquire geographic resources

TEACHER TIPS ON ACQUIRING DATA

✓ Student data is extremely confidential. In the case study, the elementary school officials were initially hesitant to release any information at all because they wanted to protect the privacy of the students. The geomatics class solved this dilemma by requesting only the minimum information necessary to complete the project: address, grade level, and a student ID. Even the ID was new and was arbitrarily assigned by the school database administrator to ensure the privacy of the student records. You may want your students to follow a similar process to avoid any problems in obtaining the necessary data.

✓ Even with the above precautions in place, it may take time to obtain the necessary permissions to release the data. Keep this in mind when planning time for working on the project. While you and your students are waiting, you may want to use the time to develop other needed data such as buffer zones for various distances from the school. Alternatively, you could also take a break from the project and come back to it once the necessary resources are in place.

Explore geographic data

- Students can create buffers either by using direct distance via the Draw Polygon tool, or if the ArcView Network Analyst extension is available, they can create service areas based on a road network. See the "On your own" section of this module for details about the ArcView Network Analyst extension (*Community Geography: GIS in Action,* page 175).

- To map the student data, students will need to geocode it first. Before geocoding the real student data, be sure that students conceptually understand the geocoding process and practice geocoding in ArcView. The module 2 exercise "Geocode crime data to map and analyze robbery hot spots" in *Community Geography: GIS in Action* presents the process of geocoding tabular data in ArcView and is a good exercise for practice.

Analyze geographic information

Students will need to analyze the data by each criterion defined by the school's transportation guidelines. In the case study and the exercise this was based on two key factors: distance and grade level. Have students complete the exercise to learn and practice one possible analysis process.

Use a critical thinking exercise to develop an analysis methodology to use for your own project. Assign students the task of listing which analysis steps should be taken and why. Remind them that they may need to query the data a number of times to cover all the defined criteria.

Act on geographic knowledge

The students should prepare a final formal presentation of their findings to the partnered school. Students should include a demonstration of the GIS analysis process in their presentation. They may also want to create an interactive product they can leave with the school administrators. Students in the case study created an ArcExplorer™ project and database for the principal so that she could interactively display and query the GIS maps when needed. An ArcView project may also be appropriate for this purpose.

The students' presentation will provide closure for the current project, while at the same time providing an opportunity for conversation about future projects. Direct feedback from the school administration can give students a sense of success and encourage them to engage in future projects. It can also let them feel appreciated as valued members of their academic community.

Other possible actions your students could take include:

- Educating elementary school students about the busing guidelines
- Hosting an open house about school transportation
- Giving a presentation to bus drivers on how GIS is used to determine who walks and who rides the bus

Module 7: Protecting the community forest

Lesson

Case study: One, two, tree—Taking a tree inventory
Exercise: Map and query a tree inventory to locate hazardous trees

Overview

In this exercise, students will explore the concept of a tree inventory and then use GIS to map and analyze one such inventory. Their task is to identify trees on a middle school property that represent a potential hazard to the school building and to overhead utility lines. First, they map the school's tree inventory by digitizing trees from an aerial photograph. Then, as they build their basemap, they create an attribute table for the digitized trees. Finally, using ArcView software's query capability, they identify the hazardous trees and create a layout to illustrate the location of those trees for the town's Public Works Department.

Estimated time

45-MINUTE CLASS	ACTIVITY
1	Lesson introduction and exercise part 1
2	Exercise part 2
3	Conclude exercise
	Begin assessment
4	Complete assessment

Materials

- *Community Geography: GIS in Action* book (one per student)
- Computers for all students (for the exercise and assessment)
- Transparency: Benefits of trees in the urban environment

Student handouts from this lesson to be photocopied:

- Assessment

Standards

GEOGRAPHY STANDARD	MIDDLE SCHOOL	HIGH SCHOOL
1 How to use maps and other geographic representations, tools, and technologies to acquire, process, and report information from a spatial perspective	The student knows and understands how to make and use maps and databases to analyze spatial distributions and patterns.	The student knows and understands how to use geographic representations and tools to analyze, explain, and solve geographic problems.
9 The characteristics, distribution, and migration of human populations on Earth's surface	The student knows and understands the effects of migration on the characteristics of places.	The student knows and understands the impact of human migration on human and physical systems.
15 How physical systems affect human systems	The student knows and understands how the characteristics of different physical environments affect human activities.	The student knows and understands how changes in the physical environment can diminish its capacity to support human activity.
16 The changes that occur in the meaning, use, distribution, and importance of resources	The student can identify and develop plans for the management and use of renewable resources.	The student knows and understands the geographic results of policies and programs for resource use and management.
18 How to apply geography to interpret the present and plan for the future	The student knows and understands how to apply the geographic point of view to solve social and environmental problems by making geographically informed decisions.	The student knows and understands how to use geographic knowledge, skills, and perspectives to analyze problems and make decisions.

Objectives

The student is able to:

- Identify benefits of trees in an urban environment
- Use GIS technology to map and analyze spatial patterns in a tree inventory
- Interpret data to identify a current problem and make a geographically informed decision about that problem

GIS skills and tools

- Zoom in on the view
- Pan the view to see different areas of the map
- Zoom to the active theme
- Open an attribute table
- Edit data in an attribute table
- Add themes to the view

- Rename a theme
- Sort the attribute table in ascending order
- Sort the attribute table in descending order
- Clear selected features
- Build a query statement
- Add text to a map layout

GEOGRAPHIC INQUIRY STEP	GIS SKILL
Ask a geographic question	• Recognize and understand geographic questions posed in a scenario.
Acquire geographic resources	• Create a basemap by digitizing feature themes from a digital orthophotograph. • Create an attribute table for digitized themes with data collected in a tree inventory.
Explore geographic data	• Sort data in the attribute table. • Observe spatial patterns of tree characteristics using the attribute table and the Query tool.
Analyze geographic information	• Visually analyze spatial patterns of tree attributes. • Create subsets of themes by using the selection features of the Query tool (new set, select from set, add to set). • Create new shapefiles and themes from subsets obtained with a query statement.
Act on geographic knowledge	• Create and print a layout to display the results of spatial analysis. • Prepare a report of tree inventory conclusions to present to appropriate municipal officials and departments.

Teacher notes

Lesson introduction

Introduce this lesson with a brainstorming activity. Ask students to take two minutes to list benefits that communities derive from the trees within their boundaries. Use the blackboard or overhead projector to list student ideas. After listing student ideas, display and discuss the transparency "Benefits of trees in the urban environment" (page 92). Identify which benefits were on the student lists and which were not.

Tell the students that they are about to read a case study about a school that undertook a project to help protect their community's trees. Instruct students to read the case study "One, two, tree—Taking a tree inventory" in *Community Geography: GIS in Action*.

Discuss the following questions:

* What is a tree inventory?
* What is the value of a tree inventory?
* Why is it important for communities to develop long-term plans for the management and care of their trees?

Exercise

Before completing this lesson with students, we recommend that you complete it on your own. Doing so will allow you to modify the activity to accommodate the specific needs of your students.

Read the exercise scenario aloud with the class. Explain that in part 1 they will use GIS to map, digitize, and explore a tree inventory. In part 2, they will query tree inventory data to identify current patterns and potential problems in an existing tree population. Clarify any questions before the students begin to work individually.

TEACHER TIPS

✓ It is not necessary for students to save their work at the end of part 1 because they will be working with a different tree theme for the remainder of the exercise.

✓ In part 2, students will need to save the shapefiles they create. Be sure to give them instructions on where they should save these new files. The software's default is to save new files in the Windows "Temp" folder on the C: drive. If students save their files here, they will not be able to find them again if they are working on a different computer.

✓ Advise students to follow the directions very carefully when working on steps that involve the Query Builder. It is essential to double-click the Fields and Values elements and single-click the Function element. If students are experiencing problems with their queries, it is usually because they did not click the elements correctly.

✓ Be sure to give students instructions on how to rename the project, save it, and record the new name so they can open it during another class or from a different computer.

Things to look for while the students are working on this activity:

* Are the students zooming in to see details on the orthophotograph (DOQ) when they digitize?
* Are the students experiencing any difficulty moving between the table and view windows?
* Are the students answering the questions as they work through the procedure?

Conclusion

Conclude the lesson by comparing student responses to question 9b in part 2 of the exercise: "List three questions about Barrington Middle School's tree population that you could answer and analyze with this data." Ask each student to read one of the questions that he or she crafted and post those questions on the blackboard or an over-head projector. As an alternative, have students write one of their questions on large sheets of paper and post these questions on the wall. When the list of questions has been generated, discuss these additional questions with your students:

* What information could the tree inventory provide that would be useful to public works officials who are responsible for maintaining the trees?
* What information could the tree inventory provide that would be useful to town officials who want to develop a long-term plan for managing the tree population?
* Would it be useful to have a tree inventory for an entire town? Would this inventory have produced the same results if it had been conducted twenty years earlier? Why or why not?

Assessment

In the middle and high school assessments, students are asked to prepare a visual presentation or printed document that advocates performing a tree inventory for an entire community.

There are a number of alternative formats for submitting the final student assessments:
- Multimedia presentation
- Illustrated brochure
- Illustrated text (word-processor) document

Middle school: Highlights skills appropriate to grades 5 through 8

Middle school students will address the following issues in their presentations:
- The benefits of trees in the urban environment
- Community changes that can endanger the tree population
- Risks of unplanned growth and urban sprawl
- The value of a tree inventory for forestry planning

High school: Highlights skills appropriate to grades 9 through 12

High school students will address the following issues in their presentations:
- The benefits of trees in the urban environment
- How reduction of the community forest affects the community's ecosystem and the ecosystems of other places
- Risks of unplanned growth and urban sprawl
- The value of a tree inventory for forestry planning

MIDDLE SCHOOL ASSESSMENT

Map and query a tree inventory to locate hazardous trees

Name _____ Date _____

After successfully completing a tree inventory of their school grounds, the Barrington students realized the value of tree inventories and wanted to persuade their town government to undertake a tree inventory for the entire community.

Assume that you are responsible for preparing a presentation to the town council with the goal of convincing them to undertake a tree inventory for the entire town. Use information from this module's case study and exercise to prepare an illustrated presentation. Your presentation should include:

• Examples of the many ways that trees improve the quality of life for community residents

• Changes that occur in settled communities over time that can have a negative effect on its tree population

• Risks of unplanned growth and development to the community and its tree population

• Map from the Barrington Middle School tree inventory illustrating the value of a tree inventory in identifying patterns in the present tree population and in developing short- and long-term goals for protecting and maintaining the tree population

MIDDLE SCHOOL ASSESSMENT RUBRIC

Map and query a tree inventory to locate hazardous trees

STANDARD	EXEMPLARY	MASTERY	INTRODUCTORY	DOES NOT MEET REQUIREMENTS
1 The student knows and understands how to make and use maps and databases to analyze spatial distributions and patterns.	Prepares one or more maps that illustrate the many benefits of conducting a tree inventory.	Prepares a map that illustrates one or more benefits of conducting a tree inventory.	Prepares a map from the tree inventory, but does not clearly illustrate the benefits of conducting a tree inventory.	Does not prepare a map from the tree inventory.
9 The student knows and understands the effects of migration on the characteristics of places.	Clearly describes changes through words and illustration that the urban forest of the community has encountered due to increased human population density over time.	Clearly describes changes that the urban forest of the community has encountered due to increased human population density over time.	Lists one or two changes to the urban forest due to increased human population density over time.	Lists change in the urban forest over time, but does not directly correlate these changes to increased human population density.
15 The student knows and understands how the characteristics of different physical environments affect human activities.	Provides several concrete examples of how the urban forest improves the quality of life for residents of the community.	Provides two to three examples of how the urban forest improves the quality of life for residents of the community.	Provides at least one example of how the urban forest improves the quality of life for residents of the community.	The student has difficulty describing at least one example of how the urban forest improves the quality of life for residents of the community.
16 The student can identify and develop plans for the management and use of renewable resources.	Identifies and explains several risks of unplanned growth and development to the community and its tree population and provides ideas for avoiding these risks.	Identifies and explains two to three risks of unplanned growth and development to the community and its tree population.	Identifies one or two risks of unplanned growth and development to the community and its tree population.	Attempts to identify at least one risk of unplanned growth and development to the community and its tree population.
18 The student knows and understands how to apply the geographic point of view to solve social and environmental problems by making geographically informed decisions.	Clearly explains the importance of the tree inventory in identifying patterns in the current urban forest and provides outlines for possible plans for future growth and maintenance.	Clearly explains the importance of the tree inventory in identifying patterns in the current urban forest and for planning for future growth and maintenance.	Identifies patterns from the tree inventory, but does not relate these patterns to planning for future growth and maintenance.	Explains the importance of tree inventories in maintenance of current trees, but does not relate current patterns to future planning.

This is a four-point rubric based on the National Standards for Geographic Education. The "Mastery" level meets the target objective for grades 5–8.

HIGH SCHOOL ASSESSMENT

Map and query a tree inventory to locate hazardous trees

Name _____ Date _____

Imagine that you are president of your school's environmental club. The club decided to try to persuade your town government to undertake a tree inventory for the entire community. As club president, you are responsible for preparing a presentation that will convince town officials of the wisdom of adopting such a policy. Although your presentation will be directed to officials in your own community, you plan to use the case study and GIS exercise from this module to support the arguments you will present. Use examples from your own community and information from this module to prepare an illustrated presentation that includes the following elements:

• Examples of the many ways that trees improve the quality of life for community residents

• Ways that reduction of the community forest over time affects the community's ecosystem and the ecosystems of other places

• Risks of unplanned growth and development to the community and its tree population (cite examples from your own and other communities)

• Map from the Barrington Middle School tree inventory illustrating the value of a tree inventory in identifying patterns in the present tree population and in developing short- and long-term goals for protecting and maintaining the tree population

HIGH SCHOOL ASSESSMENT RUBRIC

Map and query a tree inventory to locate hazardous trees

STANDARD	EXEMPLARY	MASTERY	INTRODUCTORY	DOES NOT MEET REQUIREMENTS
1 The student knows and understands how to use geographic representations and tools to analyze, explain, and solve geographic problems.	Prepares a series of maps that clearly illustrates the locations of past, current, and future trees based on findings from the Barrington tree inventory.	Prepares a map that clearly illustrates the locations of past, current, and future trees based on findings from the Barrington tree inventory.	Prepares a map that illustrates the locations of past and current trees based on findings from the Barrington tree inventory.	Prepares a map that illustrates the locations of current trees from the Barrington tree inventory.
9 The student knows and understands the impact of human migration on human and physical systems.	Describes through words and illustrations how increased human population density has affected the ecosystems of the local community as well as the ecosystems of other areas.	Clearly describes how increased human population density has affected the ecosystems of the local community as well as the ecosystems of other areas.	Describes how increased human population density has affected the local ecosystem.	Describes changes to the local ecosystem due to reduction of the urban forest, but does not relate these changes to increased human population density.
15 The student knows and understands how changes in the physical environment can diminish its capacity to support human activity.	Provides three or four examples of how the urban forest enhances the quality of human life in the community, and how its diminishment can lower the quality of life.	Provides two or three examples of how the urban forest enhances the quality of human life in the community.	Provides one or two examples of how the urban forest enhances the quality of human life.	Provides at least one example of how the urban forest beautifies the community.
16 The student knows and understands the geographic results of policies and programs for resource use and management.	Demonstrates through word and illustration, and using examples of several other communities, how unplanned growth can damage the urban forest. Compares this community to others.	Demonstrates through examples of one or two other communities how unplanned growth can damage the urban forest. Compares this community to others.	Provides an example of at least one community that has experienced damage to the urban forest caused by unplanned growth.	Does not provide examples from other communities of damage to the urban forest caused by unplanned growth.
18 The student knows and understands how to use geographic knowledge, skills, and perspectives to analyze problems and make decisions.	Explains the importance of tree inventories for community planning, establishes two or three short- and long-term goals for improvement of the urban forest, and includes a time line for the goals.	Explains the importance of tree inventories for community planning and establishes two or three short- and long-term goals for improvement of the urban forest.	Establishes one or two short- and long-term goals for improvement of the urban forest, but does not relate the tree inventory to the establishment of these goals.	Establishes one or two short-term goals for improvement of the urban forest and does not attempt to relate the tree inventory to the establishment of the goal(s).

This is a four-point rubric based on the National Standards for Geographic Education. The "Mastery" level meets the target objective for grades 9–12.

BENEFITS OF TREES IN THE URBAN ENVIRONMENT

- Reduction/Detention of surface water runoff and reduction of flood risks

- Reduction of soil erosion and sedimentation of water bodies

- Absorption of water and air pollutants

- Provision of wildlife habitat and recreational opportunity

- Storage of atmospheric carbon—countering the greenhouse effect and global warming

- Enhancement of property values

- Reduction of energy costs by providing shade and windbreaks

- Abatement and buffering of noise

- Community aesthetics and links to the past—beauty and a "sense of place"

- Psychological and sociological impacts, including lessening of stress and reduction of crime

Source: Rhode Island Urban and Community Forestry Plan. May 1999. Report Number 97. State Guide Plan Element 156, page 3.2.

On your own

Overview

This section provides guidelines and information to help you implement a similar project in your own classroom. A tree inventory and its analysis is an excellent activity with which to teach students key problem-solving skills: how to collect and organize data, how to integrate information from various sources, how to analyze patterns and relationships within data, how to collaborate with others in decision making, and how to communicate findings in a variety of formats.

Estimated time

- Four to five weeks for an individual class

This time includes introductory activities, data collection, learning GIS skills to map the inventory, inventory data analysis, final report preparation, and presentations. The various components could be spread out over an entire school year or concentrated in a focused four- to five-week time period.

Materials

- Computers for all students (one per student or student group)
- Digital orthophotograph of your school site
- Access to a color printer
- Class e-mail list for correspondence with community partners
- Reserved disk space for data storage

The students at Barrington Middle School worked in teams consisting of four students (see "Teacher tips on organizing a tree inventory with students" on page 96). The following table identifies the materials used by each team to conduct its tree inventory.

ITEM	COMMENTS
Clinometer	These are plastic "trigger-style" clinometers that can be ordered from a school laboratory equipment catalog. They measure tree height.
100-foot reel tape	Use these to measure a 100-foot distance from the base of each tree. Tree heights are calculated from that distance.
DBH tape	A DBH (diameter at breast height) tape is a forester's tool that allows students to measure diameter directly from the tree. Without this tape, they would have to calculate diameter from the tree's circumference.
Tree identification guide	You can obtain high-quality tree identification books from a variety of nonprofit organizations. Decide on one book for use by all the teams and at least two additional books for reference. See this book's Web site for more information *(www.esri.com/communitygeography)*.
Clipboard	A clipboard makes it easier to organize and write neatly on tree data collection sheets.
Metal tree tags and nails	Numbered tree tags are available from forestry suppliers. The nails are designed to do minimal damage to the tree. You need enough tags and nails for all of the trees in your inventory.

Standards

GEOGRAPHY STANDARD	MIDDLE SCHOOL	HIGH SCHOOL
1 How to use maps and other geographic representations, tools, and technologies to acquire, process, and report information from a spatial perspective	The student knows and understands how to make and use maps and databases to analyze spatial distributions and patterns.	The student knows and understands how to use geographic representations and tools to analyze, explain, and solve geographic problems.
2 How to use mental maps to organize information about people, places, and environments in a spatial context	The student knows and understands how perception influences people's mental maps and attitudes about places.	The student knows and understands how mental maps reflect the human perception of places.
4 The physical and human characteristics of places	The student knows and understands how different human groups alter places in distinctive ways.	The student knows and understands how relationships between humans and the physical environment lead to the formation of places and to a sense of personal and community identity.
6 How culture and experience influence people's perceptions of places and regions	The student knows and understands how technology affects the ways in which cultural groups perceive and use places and regions.	The student knows and understands the ways in which people's changing views of places and regions reflect cultural change.
14 How human actions modify the physical environment	The student knows and understands how human modifications of the physical environment in one place often lead to changes in other places.	The student knows and understands the ways in which technology has expanded the human capability to modify the physical environment.
15 How physical systems affect human systems	The student knows and understands how the characteristics of different physical environments affect human activities.	The student knows and understands how changes in the physical environment can diminish its capacity to support human activity.
16 The changes that occur in the meaning, use, distribution, and importance of resources	The student can identify and develop plans for the management and use of renewable resources.	The student knows and understands the geographic results of policies and programs for resource use and management.
18 How to apply geography to interpret the present and plan for the future	The student knows and understands how to apply the geographic point of view to solve social and environmental problems by making geographically informed decisions.	The student knows and understands how to use geographic knowledge, skills, and perspectives to analyze problems and make decisions.

Objectives

The student is able to:

- Digitize a project basemap from a digital orthophotograph
- Create an attribute table of data for inventoried trees
- Identify major characteristics of and problems in the community's present tree population
- Recommend actions that will facilitate the development and continuation of a healthy tree population in the community

GIS skills and tools

GEOGRAPHIC INQUIRY STEP	GIS SKILL
Ask a geographic question	• Where are the trees in your community? • In which area is it appropriate to conduct a tree inventory? • Where are the hazardous trees?
Acquire geographic resources	• Work with your local city GIS department to obtain a digital orthophotograph of the study site. • Work with local public works officials and arborists to plan a tree inventory. • Conduct a tree inventory and record data in a database that can be used in GIS software. • Create a basemap by digitizing feature themes from a digital orthophotograph.
Explore geographic data	• Sort data in the attribute table. • Observe spatial patterns of tree characteristics using the attribute table and the Query Builder.
Analyze geographic information	• Visually analyze spatial patterns of tree attributes. • Identify trees with particular characteristics by using the Query Builder to select features. • Create subsets of themes by refining the selected set (new set, select from set, add to set) with the Query Builder tool.
Act on geographic knowledge	• Create and print a layout to display the results of spatial analysis. • Prepare a report of tree inventory conclusions to present to appropriate municipal officials. • Educate others about your results.

Teacher notes

Ask a geographic question

To develop a central question for this project, begin with activities that raise awareness of the importance of trees in our lives and communities. Trees are a "green umbrella" that enhance our lives in many ways (refer to the transparency "Benefits of trees in the urban environment").

TEACHER TIPS

✓ Create a record of notable or outstanding tree specimens in your community.
✓ Conduct a poster contest on the ways that trees in your community improve the quality of life there.
✓ Invite arborists and foresters to speak to your class about the importance of trees in the urban ecosystem.
✓ Visit a tree farm, arboretum, or botanical garden.
✓ Take a walking tour around your community to observe and record the "sense of place" trees create in the town.

Once students are aware of the many benefits we derive from community trees, the next step is to identify the dangers to community trees (cutting down trees for building projects and road construction, salt and chemicals on the roads, declining soil quality, increase in impervious surfaces, air pollution, depletion of the water table, and so on).

The ultimate goal is to reach a point where students will wonder how we can protect our "green umbrella" for the future when there are so many threats to its survival. Introduce the concept of sustainable development and the underlying question of the project: "How can we manage our arboreal resources today so that future generations will continue to experience the beauty and benefits that community trees provide?"

The answer is for every community to develop short- and long-term plans for community forest management and protection. Community forestry plans begin with a tree inventory.

Now that your students understand the importance of community forestry planning and tree inventories, you must select an appropriate site for your own inventory.

TEACHER TIPS ON SITE SELECTION FOR YOUR TREE INVENTORY

✓ Choose a site that students can walk to easily.
✓ Choose a site on the school campus.
✓ Choose a park near the school campus.
✓ A site on public property is preferable because it eliminates issues of trespassing or needing to ask permission.
✓ Safety for student data collectors is paramount—choose a site that will allow you to supervise and maintain visual contact with students at all times.

Acquire geographic resources

TEACHER TIPS ON OBTAINING THE DIGITAL ORTHOPHOTOGRAPH OF YOUR STUDY SITE

✓ See the module 7 "on your own" section of *Community Geography: GIS in Action* for a discussion of possible sources of digital orthophotographs of your inventory site. By explaining that the data will be used for an educational project, you will often find that data providers are willing to waive fees for the data.

✓ Enlist the help of local GIS users to find what you need; be sure to tell them the data will be used for a school project. In some cases, data fees can be waived.

TEACHER TIPS ON ORGANIZING A TREE INVENTORY WITH STUDENTS

✓ Conduct the inventory at an appropriate time of year for your region and site. If your site has deciduous trees, be sure to conduct the inventory while there are leaves on the trees. Identifying trees without leaves is much more difficult.

✓ Introduce essential terms and equipment in the classroom. Practice using both before conducting the inventory.

✓ Hold an outdoor demonstration of inventory techniques and issues and allow practice time for student teams.

✓ Send a letter to parents requesting volunteers to help with the outdoor inventory. Having additional supervision can help, especially if your study site is a large area.

✓ Ask members of local tree societies or similar groups to serve as "on-site experts" during the inventory.

✓ Have students work in groups of four with one student designated group leader. Only the group leader should go to the teacher with problems, to verify a tree identification, or to get a new tree assignment.

✓ Have a Sign In/Sign Out sheet for equipment. This is a good job for parents. It allows you to focus on inventory questions from the students.

✓ Require students to have each tree identification verified before submitting completed data sheets. Verification should be signified by the signature or initials of a teacher or "on-site expert."

✓ Each group gets a tree data sheet with a tree ID number. They cannot get another sheet and move on to another tree until the sheet has been completed, verified, and turned in to the teacher in charge.

✓ If possible, work in blocks of time that are longer than the typical forty-five-minute class period—two-hour blocks work very well.

✓ Number the trees a day or two before taking students into the field—tree tags on school grounds are often subject to vandalism and may tend to disappear during the weeks after the numbering process.

✓ Make a site map for yourself so you know where the numbered trees are located. Printing a set of site images from your DOQ works well for this.

Explore geographic data

Technical issues

The confidence and skill level of your students will determine how you set up the GIS project for the tree inventory. Below are some options for preparing the basemap and the tree attribute table.

• Possibilities for the basemap:
  – Students work in groups, with each group responsible for digitizing their own basemap entirely.
  – Each student or group digitizes buildings, pavement, the study site outline, and wooded areas. The teacher digitizes the trees for accuracy and provides the data to all students.
  – A small team of student GIS "experts" prepares the basemap themes that the entire class will analyze.

• Possibilities for the tree attribute table:
  – Each group enters all of the data themselves.
  – The group of student GIS "experts" prepares the attribute table, either in ArcView or in a spreadsheet program, and distributes it to the class.
  – The teacher prepares a table and the students join it to their digitized tree theme.

Preliminary data exploration

• To guide preliminary exploration of the inventory data, ask students to make four observations about each of the four categories below (or other categories that may be appropriate to your site):
  – Species diversity
  – Age of trees
  – Health of trees
  – Potential problems in the tree population

- Use a projection device to demonstrate some of the techniques students can use to explore the data.
  - Sort and select features from the attribute table
  - Use the theme definition function
  - Symbolize tree data in different ways
  - Perform queries
- Compare student observations and preliminary conclusions in the classroom before beginning the final analysis. Discuss whether the conclusions are valid and supported by the data.

Analyze geographic information

It is important to give students guidelines for their analysis of the tree inventory. In Barrington, the goal of the project was to report inventory conclusions to municipal officials. Therefore, the final student project was the preparation of a summary report for town officials. They worked on their final reports in groups of four with the following instructions:

- Compare observations you made individually in your preliminary exploration of the data. Identify those that are clearly supported by the data.
- Select one observation for further exploration and analysis in each of the four (or more) categories:
  - Species diversity of the tree population in the study area
  - Age structure of the tree population in the study area
  - Health of the tree population in the study area
  - Hazardous or problem trees
- Based on your exploration, state a conclusion about each of the four (or more) categories of investigation.
- Create a map in ArcView, a graph (in ArcView or a graphing program), and a data table to support each of your conclusions.
- Identify recommendations for actions that will contribute to the long-term management and care of the trees in your inventory for each of the four categories.
- Assemble a tree inventory report including the following components:
  - An introductory paragraph explaining the importance of trees to the community
  - A paragraph defining a tree inventory and explaining its value
  - A bulleted summary of your observations and recommendations for each category
  - Maps, graphs, and tables that support those conclusions and recommendations

Act on geographic knowledge

- Find out whether you and your students can attend a school committee or town government meeting to give a presentation on your results.
- Advocate expanding the tree inventory to other parts of town.
- Invite town, county, and state officials to your school for a presentation.
- Prepare a presentation and lesson for younger children in your town and ask if you can visit their school to present it.
- Prepare an Arbor Day event for your school. Plant a tree that is an appropriate species based on your analysis.
- Prepare a tree identification map of your school grounds so that others can learn from your efforts.
- A tree inventory is an ongoing process. Your class has made the first deposit in a database of information that will continue to grow in both size and value. Prepare a presentation for younger students in your school to generate excitement and enthusiasm about continuing the work that your class has begun.
- Select a new study area and add data to the tree inventory database.
- Assist municipal officials in maintaining the database, analyzing the data, and preparing reports.

Lesson

Case study: Using site analysis to develop a wildlife area management plan
Exercise: Perform site selection for a state wildlife area

Overview

In this lesson, students will explore the variety of issues surrounding site selection and will complete an exercise to site a parking lot. They will create buffers to analyze the parking lot selection criteria and basemap data. Students will conclude the exercise by siting a third parking lot option and determining which of the three options is best.

Estimated time

45-MINUTE CLASS	ACTIVITY
1	Introduction
2	Exercise part 1
3	Exercise part 2
4	Assessment
5	Assessment

Materials

- *Community Geography: GIS in Action* book (one per student)
- Computers for all students (for the exercise and assessment)
- Large pieces of paper and markers for lesson introduction

Student handouts from this lesson to be photocopied:

- Assessment(s)

Standards

GEOGRAPHY STANDARD	MIDDLE SCHOOL	HIGH SCHOOL
1 How to use maps and other geographic representations, tools, and technologies to acquire, process, and report information from a spatial perspective	The student knows and understands how to make and use maps and databases to analyze spatial distributions and patterns.	The student knows and understands how to use geographic representations and tools to analyze, explain, and solve geographic problems.
4 The physical and human characteristics of places	The student knows and understands how to analyze the human and physical characteristics of places.	The student knows and understands the changing physical and human characteristics of places.
13 How the forces of cooperation and conflict among people influence the division and control of Earth's surface	The student knows and understands why people cooperate but also engage in conflicts to control Earth's surface.	The student knows and understands how the differing points of view and self interests play a role in conflict over territory and resources.
16 The changes that occur in the meaning, use, distribution, and importance of resources	The student can evaluate different viewpoints regarding resource use.	The student knows and understands the geographic results of policies and programs for resource use and management.
18 How to apply geography to interpret the present and plan for the future	The student knows and understands how to apply the geographic point of view to solve social and environmental problems by making geographically informed decisions.	The student knows and understands how to use geographic knowledge, skills, and perspectives to analyze problems and make decisions.

Objectives

The student is able to:

- Interpret the symbols on a topographic map to describe the physical and human characteristics of a place.
- Compare and contrast the value of maps and aerial photographs for investigating the physical and human characteristics of places.
- Describe potential risks that human activities pose in a wildlife area.
- Analyze data to determine the best locations for a parking lot and campsites.

GIS skills and tools

⊕	Zoom in on the view	▸	Select and move labels and graphics
⬇	Add data to the view	⬈	Zoom to the active theme
▦	Set scale-dependency values	▭	Select features in the view
✋	Pan the view to see different areas of the map	✗	Run a script to create a buffer
❶	Identify features on the map	▯	Clear selected features
AB	Change the font type and size of labels	▭	Draw a rectangle
▣	Change the symbol color for labels	▦	Open an attribute table
▤	Create labels	▯	Edit data in an attribute table

GEOGRAPHIC INQUIRY STEP	GIS SKILL
Ask a geographic question	• Recognize and understand geographic questions presented in a scenario.
Acquire geographic resources	• Add 100-foot setback data and an aerial photograph to the view. • Run a script to create 300- and 500-foot buffer data. • Add location data for parking lot options one and two.
Explore geographic data	• Set scale dependencies for different data layers. • Edit symbology and label data. • Explore differences and similarities between the aerial photograph and the topographic map.
Analyze geographic information	• Draw a polygon to represent the parking lot size and determine where it can be placed. • Identify a third option for the parking lot location and explain its advantages and disadvantages in a table.
Act on geographic knowledge	• Determine the best parking lot location and label it on your map. • Create and print a map layout. • Create a presentation and present this information to community groups.

Teacher notes

Lesson introduction

Divide the class into groups of two to four students. Give each group a large piece of paper and marker and present them with the following scenario:

You have just found out that your classroom will be getting new computers. The technology coordinator at your school wants you to make a proposal for where the computers should go. In the next ten minutes, make a list of all the information you need to have (and where you can get it) in order to make this proposal.

Reconvene the class and ask each student group to post their lists on the wall. A representative from each group should explain their list. Students may identify the following needs:

- How big are the computers? How much space will they take up? How many is the class going to receive? This information could come from the technology coordinator or the school principal.
- Where are the electrical outlets in the classroom? Where does the network cable come into the classroom? This information could come from a quick survey of the room or consultation with the technology coordinator.
- What are the areas that are "off limits" for a location (e.g., there could be fixed furniture like sinks or shelving)? Where are the areas that the teacher prefers to be "off limits"? This information could come from a quick survey of the room and a conversation with the classroom teacher.

Points to make during the discussion

- A thorough site selection, even for something as simple as the location of computers within a classroom, requires information from a wide variety of sources.

- Information gathered needs to be prioritized based on required criteria for the location and preferred criteria. In this example, a required criterion is the space needed by the physical size of the computer. Preferred criteria would be the desk space you would like to have next to each computer, as well as the teacher's preferences. Students may note that in some cases there are conflicting preferences.

- Connect this introductory activity to site selection in the real world by comparing information needed for computer locations with information needed for a parking lot location. Discuss similarities and differences.

- Have students read the case study "Using site analysis to develop a wildlife area management plan" in *Community Geography: GIS in Action* and tell them they will explore a variety of information sources before deciding where a parking lot should be located within a wildlife area.

- Ask students to relate what they learned about where to put computers in a classroom to the important site-selection issues identified in the case study.

Exercise

Before completing this lesson with students, we recommend that you complete it on your own. Doing so will allow you to modify the activity to accommodate the specific needs of your students.

Read the exercise scenario aloud with the class and review its goals.

TEACHER TIPS

✓ In part 1, steps 2 and 7, be sure that students understand the sometimes counterintuitive concept of small and large scale and scale dependency.

✓ Students will be saving their project as well as new shapefiles at different points in the exercise. Be sure to inform students where to save their work and how to name their files. At the end of part 1, have them record the new project name so they can open it during another class or from a different computer.

✓ If your students are using ArcView 3.1 or higher and they opt to use the Create Buffers wizard (the alternative method described on page 228), be sure they continue with the exercise at step 18.

✓ When creating the layout for their printed map at the end of the exercise, remind students to carefully evaluate which themes they want to depict on their map. You may want to coach them with tips on good cartographic design (e.g., legibility, uncluttered map, and so on) and using a map to tell a story.

Things to look for while the students are working on this activity:

- Are students able to set scale dependencies for the data?
- Do students understand the importance of the different setbacks and why they are creating buffers?
- Are students able to run the buffer script and save them as new shapefiles?
- Are students able to analyze the data to determine another parking lot location?
- Are students considering the criteria as well as other advantages and disadvantages of each potential location?

Conclusion

Divide students into groups of three to five. Conclude the lesson by having each group discuss the following questions:

- How did they decide where to put the parking lot?
- Which data sources were most useful to them in making their decision?
- What was the most difficult part of the process?
- What other things did they take into consideration when making their decision? (For example, did anyone think of traffic issues?)
- What additional data might be helpful with this type of analysis?
- What additional information would they need to determine appropriate campsite locations within the wildlife area?

Assessment

In the middle and high school assessments, students are asked to use what they learned in the exercise and additional research to site overnight camping areas in the Chuck Lewis State Wildlife Area. They will use provided criteria to guide their analysis.

Middle school: Highlights skills appropriate to grades 5 through 8

Middle school students will conduct a site analysis for three proposed campsites in the Chuck Lewis State Wildlife Area. They will prepare a map and a written report for the recreational group interested in the proposal.

High school: Highlights skills appropriate to grades 9 through 12

High school students will conduct a site analysis for three proposed campsites in the Chuck Lewis State Wildlife Area. They will prepare a map and a written report for the recreational group interested in the proposal. Additionally, they will conduct a class debate on whether overnight camping should be permitted in the wildlife area.

Divide students into three teams: those for overnight camping, those against overnight camping, and those who serve on the deciding panel. Provide students with guidelines on how the debate will be structured.

Note: This is a fictional assessment activity. In reality, the CDOW does not allow overnight camping in its state wildlife areas.

MIDDLE SCHOOL ASSESSMENT

Perform site selection for a state wildlife area

Name _____   Date _____

Because you did such a good job siting a parking lot for the Chuck Lewis State Wildlife Area, a local outdoor recreation group is seeking your help. They would like to make a proposal to the CDOW to allow overnight camping in the wildlife area and they want you to perform the site analysis. They gave you the following parameters as a guide for your analysis:

- You need to select sites for three campsites that measure 900 square feet each (30 feet × 30 feet). They must be clustered together, but can be any shape.
- The closest campsite needs to be at least 100 feet away and no more than 500 feet away from the parking lot.
- All campsites need to be at least 100 feet away from the river, but the recreation group prefers them to be at least 300 feet away from the river.
- The campsites cannot be located in the flood zone of the Yampa River.

Part 1: Select campsite locations and create a map

- Open your saved wildlife_abc.apr project and choose the parking lot with the best location. This is the parking lot you will use to select campsite locations from.
- Use what you learned in the exercise to create buffers, analyze the data, and determine where the new campsites should be located.
- Create a map that best displays the recommended campsite locations and other essential data.

Part 2: Report to the recreation group

In addition to selecting the campsite locations, the recreation group would like a document explaining your recommendations. Conduct any necessary outside research to answer the following questions. Write your answers on a separate sheet of paper.

1 Describe the advantages and disadvantages of the campsite locations you recommended.

2 If you were presenting this proposal to CDOW, how would you describe the potential negative and positive impacts of overnight camping in the Chuck Lewis State Wildlife Area? Include environmental and tourism impacts in your response.

3 What additional information or data could help you make a more informed decision about where the campsites should be located?

MIDDLE SCHOOL ASSESSMENT RUBRIC

Perform site selection for a state wildlife area

STANDARD	EXEMPLARY	MASTERY	INTRODUCTORY	DOES NOT MEET REQUIREMENTS
1 The student knows and understands how to make and use maps and databases to analyze spatial distributions and patterns.	Creates a map that displays campsite and parking lot locations. The map is symbolized appropriately and clearly communicates a message. It meets all listed requirements (e.g., appropriate themes, title, north arrow).	Creates an easy-to-read map that is symbolized appropriately, meets all listed requirements (e.g., appropriate themes, title, north arrow), and clearly identifies campsite and parking lot locations.	Creates a map that displays some campsite and parking lot locations and meets most listed requirements (e.g., appropriate themes, title, north arrow). Attempts to symbolize the map appropriately.	Creates a map that displays parking lot locations but not campsite locations. It is not symbolized appropriately or is missing several listed requirements (e.g., appropriate themes, title, north arrow).
4 The student knows and understands how to analyze the human and physical characteristics of places.	Selects the most appropriate campsite locations based on the criteria provided and additional student-defined preferences (e.g., no campsites on steep slopes or next to existing structures).	Selects the most appropriate campsite locations based on the criteria provided.	Selects campsite locations, but has difficulty meeting the criteria provided.	Does not select appropriate campsite locations or the selections are not based on the provided criteria.
13 The student knows and understands why people cooperate but also engage in conflicts to control Earth's surface.	Accurately and thoroughly describes the advantages and disadvantages of the campsite locations that are recommended. Description includes complete explanation for each point listed.	Accurately describes the advantages and disadvantages of the campsite locations that are recommended. Description includes explanation for each point listed.	Attempts to describe the advantages and disadvantages of the campsite locations that are recommended, but is incorrect on some points.	Attempts to describe the advantages and disadvantages of the campsite locations that are recommended, but is incorrect on most points.
16 The student can evaluate different viewpoints regarding resource use.	Accurately and thoroughly describes the positive and negative impacts of overnight camping. Discusses a broad range of impacts, including those to the environment, local tourism, and transportation.	Accurately describes the positive and negative impacts of overnight camping. Discusses a range of impacts that could include those to the environment, local tourism, and transportation.	Attempts to describe the positive and negative impacts of overnight camping. The list of impacts is short and brief.	Attempts to describe the positive and negative impacts of overnight camping, but is incorrect on most points.
18 The student knows and understands how to apply the geographic point of view to solve social and environmental problems by making geographically informed decisions.	Creates a varied list of the type of data and information needed to make a more informed decision about campsite locations. Explains how each piece of data or information would help the decision-making process in this geographical area.	Creates a list of the type of data and information needed to make a more informed decision about campsite locations. Explains how most data or information would help the decision-making process in this geographical area.	Creates a list of the type of data and information needed to make a more informed decision about campsite locations, but does not explain how it would help the decision-making process in this geographical area.	Creates a short list (or no list) of the data and information needed to make a more informed decision about campsite locations. Provides no explanation of how it would help the decision-making process in this geographical area.

This is a four-point rubric based on the National Standards for Geographic Education. The "Mastery" level meets the target objective for grades 5–8.

HIGH SCHOOL ASSESSMENT

Perform site selection for a state wildlife area

Name _____ Date _____

Because you did such a good job siting a parking lot for the Chuck Lewis State Wildlife Area, a local outdoor recreation group is seeking your help. They would like to make a proposal to the CDOW to allow overnight camping in the wildlife area and they want you to perform the site analysis. They gave you the following parameters as a guide for your analysis:

- You need to select sites for three campsites that measure 900 square feet each (30 feet × 30 feet). They must be clustered together, but can be any shape.
- The closest campsite needs to be at least 100 feet away and no more than 500 feet away from the parking lot.
- All campsites need to be at least 100 feet away from the river, but the recreation group prefers them to be at least 300 feet away from the river.
- The campsites cannot be located in the flood zone of the Yampa River.

Part 1: Select campsite locations and create a map

- Open your saved wildlife_abc.apr project and choose the parking lot with the best location. This is the parking lot you will use to select campsite locations from.
- Use what you learned in the exercise to create buffers, analyze the data, and determine where the new campsites should be located.
- Create a map that best displays the recommended campsite locations and other essential data.

Part 2: Report to the recreation group

In addition to selecting the campsite locations, the recreation group would like a document explaining your recommendations. Conduct any necessary outside research to answer the following questions. Write your answers on a separate sheet of paper.

1 Describe the advantages and disadvantages of the campsite locations you recommended in relation to the criteria stated above and the features on the map.

2 If you were presenting this proposal to CDOW, how would you describe the potential negative and positive impacts of overnight camping in the Chuck Lewis State Wildlife Area? Include environmental and tourism impacts in your response.

3 What additional information or data could help you make a more informed decision about where the campsites should be located?

Extension: Classroom debate

Imagine that you are at the meeting where the recreation group is making its case for overnight camping in the Chuck Lewis State Wildlife Area. Your teacher will assign you one of three roles:

- For overnight camping
- Against overnight camping
- A member of the panel who makes the final decision and provides an explanation for the decision

Prepare for the class debate by conducting research on the issue of overnight camping in recreation areas. Your teacher will provide you with specific guidelines on how the debate will be structured.

HIGH SCHOOL ASSESSMENT RUBRIC

Perform site selection for a state wildlife area

STANDARD	EXEMPLARY	MASTERY	INTRODUCTORY	DOES NOT MEET REQUIREMENTS
1 The student knows and understands how to use geographic representations and tools to analyze, explain, and solve geographic problems.	Creates a map that displays campsite and parking lot locations. The map is symbolized appropriately and clearly communicates a message. It meets all listed requirements (e.g., appropriate themes, title, north arrow).	Creates an easy-to-read map that is symbolized appropriately, meets all listed requirements (e.g., appropriate themes, title, north arrow), and clearly identifies campsite and parking lot locations.	Creates a map that displays some campsite and parking lot locations and meets most listed requirements (e.g., appropriate themes, title, north arrow). Attempts to symbolize the map appropriately.	Creates a map that displays parking lot locations, but not campsite locations. It is not symbolized appropriately or is missing several listed requirements (e.g., appropriate themes, title, north arrow).
4 The student knows and understands the changing physical and human characteristics of places.	Selects the most appropriate campsite locations based on the criteria provided and additional student-defined preferences (e.g., no campsites on steep slopes or next to existing structures).	Selects the most appropriate campsite locations based on the criteria provided.	Selects campsite locations, but has difficulty meeting the criteria provided.	Does not select appropriate campsite locations or the selections are not based on the provided criteria.
13 The student knows and understands how differing points of view and self interests play a role in conflict over territory and resources.	Accurately and thoroughly describes the advantages and disadvantages of the campsite locations that are recommended. Description includes complete explanation for each point listed. Able to effectively assume the role assigned during the class debate.	Accurately describes the advantages and disadvantages of the campsite locations that are recommended. Description includes explanation for each point listed. Able to assume the role assigned during the class debate.	Attempts to describe the advantages and disadvantages of the campsite locations that are recommended, but is incorrect on some points. Has difficulty assuming the role assigned during the class debate.	Attempts to describe the advantages and disadvantages of the campsite locations that are recommended, but is incorrect on most points. Unable to assume the role assigned during the class debate.
16 The student knows and understands the geographic results of policies and programs for resource use and management.	Accurately and thoroughly describes the positive and negative impacts of overnight camping. Discusses a broad range of impacts, including those to the environment, local tourism, and transportation, resulting from a decision to allow overnight camping.	Accurately describes the positive and negative impacts of overnight camping. Discusses a range of impacts that could include those to the environment, local tourism, and transportation, resulting from a decision to allow overnight camping.	Attempts to describe the positive and negative impacts of overnight camping. The list of impacts is short and brief.	Attempts to describe the positive and negative impacts of overnight camping, but is incorrect on most points.
18 The student knows and understands how to use geographic knowledge, skills, and perspectives to analyze problems and make decisions.	Creates a varied list of the type of data and information needed to make a more informed decision about campsite locations. Explains how each piece of data or information would help the decision-making process in this geographical area. Provides logical, thorough, and persuasive arguments in the class debate.	Creates a list of the type of data and information needed to make a more informed decision about campsite locations. Explains how most data or information would help the decision-making process in this geographical area. Provides logical and persuasive arguments in the class debate.	Creates a list of the type of data and information needed to make a more informed decision about campsite locations, but does not explain how it would help the decision-making process in this geographical area. Provides some persuasive arguments in the class debate.	Creates a short list (or no list) of the data and information needed to make a more informed decision about campsite locations. Provides no explanation of how it would help the decision-making process in this geographical area. Provides few, if any, persuasive arguments in the class debate.

This is a four-point rubric based on the National Standards for Geographic Education. The "Mastery" level meets the target objective for grades 9–12.

On your own

Overview

This section provides guidelines and information to help you implement a similar project in your own classroom. A site analysis is an excellent activity with which to teach students key problem-solving skills: how to collect and organize data, how to integrate information from various sources, how to analyze patterns and relationships within data, how to collaborate with others in decision making, and how to communicate findings in a variety of formats.

Estimated time

The time required for a site analysis project will depend largely on what you are siting and who you are working with on the project. Parking lot site analysis was one component of many the FLITE students tackled throughout an entire school year.

If you are just starting out with community-mapping projects, choose a project at your school that will take a week or two to complete with your students. As you and your students gain more confidence and experience with GIS, expand the project to include community partners.

Materials

- Computers for all students (one per student or student group)
- Class e-mail list for correspondence with community partners
- Color printer for maps

The FLITE students worked in teams of two or more students. The following table identifies the materials used by two or more teams during field data collection and during the presentation phases:

ITEM	COMMENTS
GPS units	These can be borrowed from community partners or purchased with an educational discount. Make sure each field collection team has a GPS unit with which to take readings.
Digital camera	These can be borrowed from community partners or purchased with an educational discount. Groups can share a digital camera with other groups. Photographs are very helpful to reference in site analysis.
Vehicles in which to transport students	Ideally, your data collection location is within walking distance of school. If it is not, consider transporting students in small groups using school vans. Otherwise, you may spend a lot of money on bus transportation.
Computer, projector, and large screen	Computers are necessary for GIS analysis. Each student group should have access to one computer. A projector and large screen are useful for sharing information within class and later for presenting information to the public.
Photocopies for public outreach	Budget time and money for photocopying. You will need to send letters to parents and students may correspond with community partners. You will also create informational brochures, announcements, and flyers to distribute to the public.

Standards

GEOGRAPHY STANDARD	MIDDLE SCHOOL	HIGH SCHOOL
1 How to use maps and other geographic representations, tools, and technologies to acquire, process, and report information from a spatial perspective	The student knows and understands how to make and use maps and databases to analyze spatial distributions and patterns.	The student knows and understands how to use geographic representations and tools to analyze, explain, and solve geographic problems.
4 The physical and human characteristics of places	The student knows and understands how to analyze the human and physical characteristics of places.	The student knows and understands the changing physical and human characteristics of places.
6 How culture and experience influence people's perception of places and regions	The student knows and understands how personal characteristics, culture, and technology affect perception of places and regions.	The student knows and understands why different groups of people within a society view places and regions differently.
13 How the forces of cooperation and conflict among people influence the division and control of Earth's surface	The student knows and understands why people cooperate but also engage in conflicts to control Earth's surface.	The student knows and understands how the differing points of view and self interests play a role in conflict over territory and resources.
14 How human actions modify the physical environment	The student knows and understands how human modifications of the physical environment in one place often lead to changes in other places.	The student knows and understands the significance of the global impacts of human modification of the physical environment.
15 How physical systems affect human systems	The student knows and understands how the characteristics of different physical environments provide opportunities for or place constraints on human activities.	The student knows and understands strategies to respond to constraints placed on human systems by the physical environment.
16 The changes that occur in the meaning, use, distribution, and importance of resources	The student can evaluate different viewpoints regarding resource use.	The student knows and understands the geographic results of policies and programs for resource use and management.
18 How to apply geography to interpret the present and plan for the future	The student knows and understands how to apply the geographic point of view to solve social and environmental problems by making geographically informed decisions.	The student knows and understands how to use geographic knowledge, skills, and perspectives to analyze problems and make decisions.

Objectives

The student is able to:

- Create a map from tabular and spatial data
- Create buffer zones defined by distance
- Work with community partners to create maps illustrating the spatial interpretation of planning and zoning guidelines
- Analyze possible site options in relation to buffer zones
- Identify and create maps of possible sites and present those options in a public forum

GIS skills and tools

GEOGRAPHIC INQUIRY STEP	GIS SKILL
Ask a geographic question	• What size parking lot or facility is needed? • What areas should be avoided? • Where should the parking lot be in relation to the road and other buildings? • Where is the access for the parking lot? • How do county or agency planning and zoning policies influence parking lot locations in this area?
Acquire geographic resources	• Obtain basemap information of the local area (roads, rivers, land use, boundaries, landmarks, parcel information, etc.). • Work with a local GIS specialist to obtain digital quadrangles and aerial photographs and other feature maps from the city and county. • Collect data in the field: location of human-made structures on the property, ditches, roads, and natural features. • Record data in a table that can be imported into ArcView.
Explore geographic data	• Create buffers around features where the parking lot cannot be located. Refer to local planning and zoning laws for specifics. • Visually analyze the available space and how it conflicts with other human-made structures that cannot be removed. • Determine whether you need additional data for analysis. Get it if necessary.
Analyze geographic information	• Within ArcView, create a rectangle that represents the size of the parking lot. • Identify probable locations based on buffers, preexisting structures, proximity to the road, and safety. • Create maps that display options for parking lot locations.
Act on geographic knowledge	• Summarize your recommendations. • Prepare presentations and practice them. • Give presentation to appropriate decision-making panel (e.g., your town's planning and zoning commission or school board). • Invite local media to your presentation. • Educate others about your recommendations by creating and distributing informational brochures and posters. • Develop a plan to follow up with the project after the parking lot has been built. • Explore ways to use GIS to solve other problems in your community.

Teacher notes

Ask a geographic question

With your students, brainstorm ideas on site selection activities in your community. One place to start is with your own school. Are there plans for the school to add more parking or construct a new athletic field or classroom building? If so, your students could be very helpful in selecting a site. If your school is not discussing expansion options, have your students consult the local newspaper for ideas.

Another way to learn about possible site selection projects is to consult local or state agencies or community groups. In many cases, these organizations have long lists of projects with few resources to complete them. They usually welcome any help and can provide significant guidance. The FLITE students at Steamboat Springs High School helped the Colorado Division of Wildlife (CDOW) draft a management plan for a new wildlife area much more quickly than if CDOW approached the project on its own.

When partnering with a community organization, be sure they are interested and willing to work with students. Identify the roles the students and teacher will play in the project and those the community partner will play. For example, the community partner may provide GPS units and instruction on how to use them. Students may collect the data in the field and analyze it. Both the teacher and the community partner should define the needs and goals of the project and plan the project together.

TEACHER TIPS ON PROJECT PLANNING

✓ Educate partner(s) about the curricular needs of the project, as well as the academic schedule or time line. Tailor the project accordingly.

✓ Train teachers in GIS. Have them take a course online, sign up for a teacher training institute, or attend a one-day workshop specific to the project.

✓ Set up planning time for all the teachers involved. Typically, this would be most useful before school starts so the project can be woven into the curriculum. Establish a time line and plan project-related events and activities.

✓ Schedule class time for partners and community mentors to meet with students. This way, students can prepare questions in advance and mentors can create time in their schedules to meet with students.

✓ Allow students to play a large role in structuring the strategy for solving the community problem.

✓ Inform parents about the project through a letter. Include information about what is expected of the students and any special dates or events the students will be required to attend, and encourage or solicit parent involvement.

Acquire geographic resources

When performing a site analysis, you will need to acquire a wide variety of data. Even though the data will come from various sources, it can be organized into two main groups: digital data files and data that you and your students will collect in the field.

TEACHER TIPS ON ACQUIRING DIGITAL FILES

✓ Consult with local city and county GIS departments for local data. Often, they will supply data to educators free of charge or for a nominal fee. Seek them out for all local basemap data.

✓ Get metadata for all your data, and be sure you understand what the use constraints are and how your data source wants to be acknowledged in any presentations or articles. For more information on metadata, refer to "On your own: Project planning," pages 248 and 250 of *Community Geography: GIS in Action*.

✓ Discuss data standards with the community partner(s) and make sure the data you obtain meets those standards (e.g., map projection, units, geographic area).

TEACHER TIPS ON DATA COLLECTION WITH STUDENTS

✓ Set up GPS data dictionaries before going into the field.

✓ Save money by borrowing GPS units or other data collection equipment from a local partner.

✓ Establish a data collection protocol and make sure it matches what the community partner expects (units, accuracy, attributes, and so on).

✓ Conduct practice sessions with students before collecting the GPS data for the project.

✓ If conducting a survey is part of your data collection, create a survey that generates results that can be tallied into a data table that can be imported into ArcView. This way, you can overlay your survey data with other project data for analysis.

✓ Solicit the help of parents or other teachers to serve as chaperones when students are collecting data in the field. Be sure to assign each adult a specific job like distributing field equipment or helping a student team conduct their tests.

✓ Be sure students are prepared with the right type of attire for the area in which they will be collecting data. For example, if it is rugged terrain, make sure students wear hiking boots, bring water, and dress accordingly.

✓ If appropriate, help students create well-designed public opinion surveys as part of their data collection.

Explore geographic data

Technical issues

The confidence and skill level of your students will determine how you set up the GIS project. Some options for preparing the basemap are as follows:

- Students work in groups, with each group entirely responsible for creating its own basemap.
- Each student or group digitizes buildings or other features, the study site outline, and roads.
- A small team of student GIS "experts" prepares the basemap themes that the entire class will analyze.

When you return to the classroom after field data collection, you will have multiple data-recording sheets. It is necessary to aggregate that information into one form for students to enter. Set up a data table on the whiteboard or an overhead projector, where you or a student records the data collected.

Preliminary data exploration

- To guide preliminary exploration of the field data and the basemap data, ask students to make observations about the spatial patterns they notice.
- Use a projection device to demonstrate some of the techniques students can use to explore the data:
  - Sort and select features from the attribute table.
  - Reorder the table of contents.
  - Symbolize data in different ways.
  - Perform queries.
  - Visually compare themes to discover feature relationships (e.g., river buffer versus flood-zone theme; older topographic map versus recent aerial image).
- Compare student observations and preliminary conclusions in the classroom before beginning the site analysis. Discuss whether the conclusions are valid and supported by the data.

Prepare a progress report that each student team must fill out periodically. It should include what they have accomplished during a specific time period and what work they need to do in an upcoming time period. Require that they submit time lines for their work, a list of types of support they will need (e.g., mentor/teacher or specific supplies and data), and a general statement of how things are going for the group.

Analyze geographic information

- Have students perform complete research on the requirements that would be necessary for the site selection. For example, they should be aware of local planning and zoning laws regarding the location of parking lots or the protection of animal and plant habitats.
- Help your students list the criteria that will be used. Which criteria are requirements and which are preferred? Categorize the criteria by type. For example, will you evaluate certain ones using a buffer of specified distance, and others through spatial overlay (theme-on-theme selection)?
- Help students plan the procedure they will use to select the site options. Depending on the specific nature of your project, help them understand or research alternative methodologies.
- If the analysis techniques are complex or lengthy, you may want to help students practice the skills that will be needed for the analysis. For example, you could break the analysis down into several smaller exercises or use a prototype study area to practice on.

Act on geographic knowledge

After your students complete their analysis, it is time to act on their findings. This is a critical step in the geographic inquiry process that is often overlooked. Be sure to build time into your schedule for students to do any of the following:

- Create presentation layouts that visually display the entire area and your chosen sites. Use different maps to highlight the pros and cons of each possible site.
- Conduct presentations for your community partner and the public for decision making. Be sure students describe their process of site selection and why each site was chosen.
- Write press releases announcing your presentations and invite local media resources, school officials, and parents.
- Have students write an article for the school newspaper or Web site.
- Give the data and project to your community partner so they have this information for future decision making. Keep a copy at school for future GIS projects.
- Educate students in lower grades about the power of GIS to solve community problems by making presentations to their classes.
- Conduct public presentations or write articles for local newspapers and magazines to educate the general public about your study.

References and resources

GENERAL RESOURCES

Envirolink Network. Educational resources by topic. Multiple resources on over twenty-five topics. www.envirolink.org

ESRI. ESRI Schools and Libraries Program. Information about using GIS in K–12 schools, public or college libraries, museums, nature centers, and other sites for educating the public. www.esri.com/k-12

Foresman, Joyce, and United Nations Environment Programme. 2002. *My community, our earth: A student project guide to sustainable development and geography.* Redlands, Calif.: ESRI Press.

GLOBE Program. A worldwide, hands-on, primary and secondary school-based education and science program. Download data and lessons. www.globe.gov

Haas, Tony, and Paul Nachtigal. 1998. *Place value: An educator's guide to good literature on rural lifeways, environments, and purposes of education.* Charleston, W.V.: Clearinghouse on Rural Education and Small Schools.

International Society for Technology in Education. 2000. *National educational technology standards for students: Connecting curriculum and technology.* Eugene, Oreg.: National Educational Technology Standards Project.

KanGIS. PathFinder Science. GIS for Schools. Links to valuable resources and professional development opportunities. kangis.org

Malone, Lyn, Anita M. Palmer, and Christine L. Voigt. 2002. *Mapping our world: GIS lessons for educators.* Redlands, Calif.: ESRI Press.

May, Stuart, and Tony Thomas. 1994. *Fieldwork in action 3: Managing out-of-classroom activities.* Indiana, Penn.: National Council for Geographic Education.

National Audubon Society. *National Audubon Society field guide to North American trees: Eastern region.* 1980. New York: Alfred A. Knopf.

National Geographic Society. 1994. *The national geography standards, 1994.* Washington, D.C.: The Geography Education Standards Project.

National Geographic Society Foundation. Explore resources for teachers and funding opportunities for classroom projects. www.nationalgeographic.org/foundation

National Research Council—National Academy of Sciences. 1995. *National science education standards.* Washington, D.C.: National Academy Press.

National Science Foundation. Learn about funding opportunities for classroom projects. www.nsf.gov

Orion Society. Explore Stories in the Land Teaching Fellowships and other place-based education initiatives. www.orionsociety.org

Petrides, George A. 1958. *A field guide to trees and shrubs.* Peterson Field Guides. Boston, New York: Houghton Mifflin Company.

Rice, G. A., and T. L. Bulman. 2001. *Fieldwork in the geography curriculum: Filling the rhetoric–reality gap.* Pathways in Geography Series, No. 22. Indiana, Penn.: National Council for Geographic Education.

SchoolGrants. Search by geography or topic for organizations offering grants to educators. www.schoolgrants.org

Sobel, David. 1998. *Mapmaking with children: Sense of place education for the elementary years.* Portsmouth, N.H.: Heineman.

The Earth Science Educator. Lessons, activities, simulations, and many more valuable links and resources. esdcd.gsfc.nasa.gov/esd/edu

U.S. Department of the Interior. U.S. Geological Survey. Earth Science Information Center. ask.usgs.gov

U.S. Environmental Protection Agency. Enviromapper. Links to online mapping sites for a number of environmental issues. www.epa.gov/enviro/enviromapper.html

Woodrow Wilson National Fellowship Program. Explore institute opportunities and the 1997 Environmental Institute on GIS and land use. www.woodrow.org/teachers

MODULE 1: GIS BASICS

Audet, Richard, and Gail Ludwig. 2000. *GIS in schools.* Redlands, Calif.: ESRI Press.

ESRI. 1999. *Getting to know ArcView GIS.* 3d ed. Redlands, Calif.: ESRI Press.

ESRI. ESRI Schools and Libraries Program. Information about using GIS in K–12 schools, public or college libraries, museums, nature centers, and other sites for educating the public. www.esri.com/k-12

ESRI–Canada. ESRI–Canada Schools and Libraries Program. Download a free copy of the School Tools ArcView extension. k12.esricanada.com

Gersmehl, Philip J. 1991. *The language of maps.* Pathways in Geography Series, No. 1. Indiana, Penn.: National Council for Geographic Education.

Monmonier, Mark. 1993. *Mapping it out: Expository cartography for the humanities and social sciences.* Chicago: University of Chicago Press.

ProTeacher. Click on Social Studies/Geography/Map Skills for lessons on mapping techniques. www.proteacher.com

U.S. Department of the Interior. U.S. Geological Survey. Comprehensive resources for parents, teachers, and students. Explore lessons on the Learning Web. mapping.usgs.gov

MODULE 2: REDUCING CRIME

National Crime Prevention Council. Links to resources for youth and teenagers about crime-prevention subtopics. www.ncpc.org

National Teens, Crime, and the Community. Contains information and resources for both teens and adults in becoming active in crime prevention at the local level. www.nationaltcc.org

MODULE 3: A WAR ON WEEDS

California Environmental Resources Evaluation System. Web site includes California Resource Agency Funding Matrix for Northern California. ceres.ca.gov

Missoula County Conservation District. *Montana weed project teachers handbook.* 3550 Mullan Road, Suite 106, Missoula, MT 59808-5125.

Missoula County Conservation District. War on Weeds Teacher Curriculum. Retrieved December 2002. mtwow.org/teacher-curriculum.htm

Taylor, Ronald J. 1990. *Northwest weeds: The ugly and beautiful villains of fields, gardens, and roadsides.* ISBN 0-87842-249-8, Item No. CQ417 (paperback). Missoula, Mont.: Mountain Press Publishing Company, Inc.

U.S. Department of Agriculture. National Agricultural Library. Federal and state invasive species activities and programs. Contains a comprehensive list of federal and private organizations with an interest in the prevention, control, or eradication of invasive species. www.invasivespecies.gov

U.S. Department of the Interior. Bureau of Land Management. What's Wrong With This Picture: Invasive Weeds: A Growing Pain. Retrieved October 2002. www.blm.gov/education/weed/weed.html

MODULE 4: TRACKING WATER QUALITY

Bridging the Watershed. Valuable resources for teachers and students. Learn how to identify plants and macroinvertebrates and to follow an anadromous fish in its migration. www.bridgingthewatershed.org/students.html

Community Learning Network. Approximately forty lessons or units devoted to water quality. www.cln.org

Earth Day Network. Explore the Teacher's Corner for lesson plans and ideas. www.earthday.net

Global Rivers Environmental Education Network. GREEN Hands-on Center. Web site for learning about water quality and watersheds. www.earthforce.org/green

Johnson, Robert L., Scott Holman, and Dan Holmquist. 2000. *Water quality with computers: Using Logger Pro: Water quality tests using Vernier sensors.* Vernier Software and Technology, 13979 S.W. Millikan Way, Beaverton, OR 97005-2886.

Kesselheim, A. S. 1998. *WOW! The wonders of wetlands: An educator's guide.* Environmental Concern, Inc., and The Watercourse funded by the Environmental Protection Agency, Region VIII, and the U.S. Department of the Interior, Bureau of Reclamation. St. Michaels, Md.

National Geographic Society. Geography Action! Rivers 2001. Poster and Web site. www.nationalgeographic.com/geographyaction

North Dakota State University Extension Service. *Look out below: A groundwater guidebook for youth.* 1994. Fargo, N.D.

U.S. Department of Agriculture. Agricultural Research Service. National Agricultural Library Water Quality Information Center. Click on Water and Agriculture and then Funding Sources for Water Quality. Fifty links to various funding sources for water-quality research and testing. www.nal.usda.gov/wqic/#1

U.S. Department of Agriculture. National Agricultural Library. Gateway to federal and state invasive species activities and programs. Both water and noxious weed resources. Retrieved December 2002. www.invasivespecies.gov

U.S. Department of the Interior. U.S. Geological Survey. Water education resources. A comprehensive page of thirty links to education lessons and activities. water.usgs.gov/education.html

Water Environment Research Foundation. Click on funding opportunities: Available Funding, Collaborations, Endowments. www.werf.org

MODULE 5: INVESTIGATING POINT-SOURCE POLLUTION

Benhart, John E., and Alex R. Margin Jr. 1994. *Wetlands: Science, politics, and geographical relationships.* Pathways in Geography Series, No. 20. Indiana, Penn.: National Council for Geographic Education.

Byrne, Arthur. 2001. *Pollution and its management.* Newtonards, Northern Ireland: Colourpoint Books.

Nine Mile Run Greenway Project. Education Modules and Teaching Materials for Teachers: Urban Watersheds. Classroom and field activities, glossary, and downloadable guide for fieldwork. slaggarden.cfa.cmu.edu/education/modules

Pennsylvania Department of Environmental Protection. Explore the Educators site for lesson plans and hot topics on various environmental issues, including recycling. www.dep.state.pa.us/educators

Waterstone, Marvin. 1992. *Water in the global environment.* Pathways in Geography Series, No. 30. Indiana, Penn.: National Council for Geographic Education.

MODULE 6: GETTING KIDS TO SCHOOL

U.S. Department of Transportation. Federal Highway Administration. Federal Highway Administration Education Pages. www.fhwa.dot.gov/education

U.S. Department of Transportation. Federal Transit Administration. Transit City, USA. Education Web site by the Federal Transit Administration. www.fta.dot.gov/transcity

MODULE 7: PROTECTING THE COMMUNITY FOREST

American Forest Foundation. Project Learning Tree. A K–12 education program for students and teachers. www.plt.org

Athenic Systems. TreeGuide. Locate tree information using botanical or common tree name. www.treeguide.com

Judith H. Blau, Inc. Treetures Web site. Colorful pages and learning activities for younger students. Copyright 2002. www.treetures.com/Welcome.html

May, Stuart, and Tony Thomas. 1994. *Fieldwork in action 3: Managing out-of-classroom activities.* Indiana, Penn.: National Council for Geographic Education.

National Audubon Society. *National Audubon Society field guide to North American trees: Eastern region.* 1980. New York: Alfred A. Knopf.

New Hampshire Space Grant Consortium. Forest Watch: Students and Scientists Working Together Determining the Health of New England Forests. Regional interactive project focusing on white pine as an indicator of environmental change. www.forestwatch.sr.unh.edu

Petrides, George A. 1958. *A field guide to trees and shrubs.* Peterson Field Guides. Boston, New York: Houghton Mifflin Company.

Rice, G. A., and T. L. Bulman. 2001. *Fieldwork in the geography curriculum: Filling the rhetoric–reality gap.* Pathways in Geography Series, No. 22. Indiana, Penn.: National Council for Geographic Education.

State of Rhode Island Department of Environmental Management. Bureau of Natural Resources. Division of Forest Environment. Rhode Island Urban and Community Forestry Plan. May 1999. Report Number 97. State Guide Plan Element 156. Page 3.2.

The Arbor Day Foundation. Educational activities and resources, including an online tree identification book. www.arborday.org

MODULE 8: SELECTING THE RIGHT LOCATION

Eisenhower National Clearinghouse for Science and Mathematics Education. Extensive listing of curriculum resources, professional resources, and Web links for teachers. www.enc.org

Lieberman, Gerald A., and Linda L. Hoody. 1998. *Closing the achievement gap: Using the environment as an integrating context for learning.* Poway, Calif.: Science Wizards. 619-676-0272.

Orr, David W. 1994. *Earth in mind: On education, environment, and the human prospect.* Washington, D.C.; Covelo, Calif.: Island Press.

NATIONAL GEOGRAPHY STANDARDS

STANDARD	MODULE 1	MODULE 2	MODULE 3	MODULE 4	MODULE 5	MODULE 6	MODULE 7	MODULE 8
1 How to use maps and other geographic representations, tools, and technologies to acquire, process, and report information from a spatial perspective	★	★ ●	●	●	★ ●	★ ●	★ ●	★ ●
2 How to use mental maps to organize information about people, places, and environments in a spatial context		●					●	
3 How to analyze the spatial organization of people, places, and environments on Earth's surface	★	★ ●	●	★ ●	●	★ ●		
4 The physical and human characteristics of places	★		★ ●	●	★ ●		●	★ ●
5 That people create regions to interpret Earth's complexity		★ ●				★ ●		
6 How culture and experience influence people's perceptions of places and regions		●					●	●
7 The physical processes that shape the patterns of Earth's surface				●				
8 The characteristics and spatial distribution of ecosystems on Earth's surface			★ ●	★ ●				
9 The characteristics, distribution, and migration of human populations on Earth's surface							★	
10 The characteristics, distribution, and complexity of Earth's cultural mosaics								
11 The patterns and networks of economic interdependence on Earth's surface								
12 The processes, patterns, and functions of human settlement					●			
13 How the forces of cooperation and conflict among people influence the division and control of Earth's surface								★ ●
14 How human actions modify the physical environment			★ ●	●	★ ●		●	●
15 How physical systems affect human systems			●	●			★ ●	●
16 The changes that occur in the meaning, use, distribution, and importance of resources							★ ●	★ ●
17 How to apply geography to interpret the past					●			
18 How to apply geography to interpret the present and plan for the future		★ ●	★ ●	★ ●	★ ●	★ ●	★ ●	★ ●

★ All standards addressed in each lesson are assessed in each lesson's assessment.

● The standards listed for each *On your own* project are likely to be addressed if you did a project in your own community. Specific standards will vary based on the individual characteristics of each community GIS project.

NATIONAL SCIENCE AND TECHNOLOGY STANDARDS

	Science standards									Technology standards					
STANDARD	A	B	C	D	E	F	G	H	1	2	3	4	5	6	
MODULE 1	★ ●	★ ●							★ ●	★ ●	★ ●	●	★ ●	★ ●	
MODULE 2	★ ●						●		★ ●	★ ●	★ ●	★ ●	★ ●	★ ●	
MODULE 3	★ ●			★ ●			★ ●		★ ●	★ ●	★ ●	●	★ ●	★ ●	
MODULE 4	★ ●		★ ●	★ ●			●		★ ●	★ ●	★ ●	●	★ ●	★ ●	
MODULE 5	★ ●			●			★ ●		★ ●	★ ●	★ ●	★ ●	★ ●	★ ●	
MODULE 6	★ ●						●		★ ●	★ ●	★ ●	★ ●	★ ●	★ ●	
MODULE 7	★ ●			★ ●			★ ●		★ ●	★ ●	★ ●	★ ●	★ ●	★ ●	
MODULE 8	★ ●			●		●	●		★ ●	★ ●	★ ●	●	★ ●	★ ●	

A Unifying concepts and processes in science
B Science as inquiry
C Physical science
D Life science
E Earth and space science
F Science and technology
G Science in personal and social perspectives
H History and nature of science

1 Basic operations and concepts
2 Social, ethical, and human issues
3 Technology productivity tools
4 Technology communications tools
5 Technology research tools
6 Technology problem-solving and decision-making tools

★ These standards are met by the GIS exercise and lesson assessment for middle and high school students.

● The standards listed for each *On your own* project are likely to be addressed if you did a project in your own community. Specific standards will vary based on the individual characteristics of each community GIS project.

MODULE 1: GIS BASICS
Exercise: Explore and label community features data for a city visitors map

3a	What colors are the streets on your map?	Blue, dark blue, red
3b	What business is represented by the eyeglasses symbol?	Krikorian Theater
3c	What landmark is next to the A. K. Smiley Library?	Post office
4a	What type of landmark is ESRI, Inc.?	Software company
5a	What type of landmark is the A. K. Smiley Library?	Library and shrine
5b	What is the address of the A. K. Smiley Library?	125 W. Vine St.
5c	Why is this location on the map important to know?	Those requesting the map are visiting Redlands to see the Abraham Lincoln Memorial Shrine, which is part of the A. K. Smiley Library.
8a	What happened to the map when you turned on Lodging.shp?	Colored dots appeared on the map.
8b	Does the Coffee_dessert.shp theme appear on your map?	No
8c	Why or why not?	The theme needs to be turned on to see it on the map.
11a	What change did you see on the map?	The dots are replaced with a blue roadside sign that represents lodging.
13a	Where does the new title Coffee and Dessert Shops appear on your screen?	It replaces Coffee_dessert.shp in the table of contents.
13b	Why is it sometimes helpful to change the theme name from the file name to something else?	File names are typically short and do not accurately reflect the content of the data. When a change is made to a theme name, it is displayed in the table of contents and in the legend on the printed map. Descriptive theme names help people unfamiliar with data-naming conventions understand the map.
16a	What happened to the map?	The Best Western Inn symbol turned yellow.
16b	What Redlands landmark is close to the Best Western?	ESRI
17a	What is the name and address of the hotel you selected?	Thrift-T-Lodge, 511 E. Redlands Blvd.

MODULE 2: REDUCING CRIME
Exercise: Geocode crime data to map and analyze robbery hot spots

PART 1

5a Which field contains the road's name?

Fname

5b Which field tells you if the road is called a street, avenue, lane, and so on?

Ftype

5c How many fields in the list contain ZIP Code information?

Two

8a Which five pieces of information must ArcView have to geocode according to the US Streets style?

1. LeftFrom (Left from address)
2. LeftTo (Left to address)
3. RightFrom (Right from address)
4. RightTo (Right to address)
5. Street Name (Street name)

10a What happens?

A large black dot appears on the map at the location of Bishop Dunne Catholic School.

13a How many crimes are recorded in the table?

98

13b Indicate what information is contained in the table by writing the field names (or "not included").

Information	Field name
Address where the crime took place	Evtloca
ZIP Code where the crime took place	Not included
Type of crime	Offense
Watch	Watch
Time of day	Eevttime
Police officer on duty	Not included
Beat number	Beat

19a How many addresses found a good match?

89

19b What percentage of the addresses is this?

91 percent

22a Is the building number (8081) within the address ranges for this street segment?

Yes

22b Does the street name match exactly?

Yes

22c What information does the candidate include that is not contained in the address?

W (West) or suffix direction

26a What patterns do you see in the distribution of robberies?

Possible observations include:

- Most robberies occur on the east and northeast side.
- No robbery in the western section.
- Robberies occur along arterials.
- Cluster of robberies in the northeast section.

26b Do most robberies occur along major streets (arterials) or away from them?

Most robberies occur along arterials.

PART 2

19a How does the pattern of business robberies in the southwest division differ from that of robberies of individuals?

Business robberies tend to be located in the northeastern part of the map while individual robberies are spread out across the map. Also, business robberies are located along arterials while individual robberies happen mostly on the smaller roads.

25a In general, are certain types of robbery more prevalent in some hot spots than in others? Describe your observations.

Student answers will vary depending on the location and size of the robbery hot spots they create. In general, students will find that individual robberies are most prevalent in most hot spots; occasionally they will create a hot spot where residential robberies or business robberies are more numerous. However, the relative prevalence of residential or business robberies may vary significantly from one hot spot to another.

26a What is the relationship of robberies to arterial streets in the hot spots you identified? For example, does a string of similar robberies occur along a particular arterial?

Student answers will vary. Possible answers include:
- In all hot spots, at least one type of robbery occurred along an arterial.
- In hot spot 1, all business robberies took place on the same arterial.
- Several robberies occurred along an arterial but they were a mixture of robbery types.

32a, Student answers will vary. The following table
33a, contains sample answers.
34a

Hot-spot results table

Hot spot	IDs of beats in hot spot	Business robberies	Individual robberies	Residential robberies	Total robberies
1	411, 418, 421, 422, 425	3	10	5	18
2	412, 415, 416, 419, 441	3	13	3	19
3	442, 443, 444, 447, 453	2	6	1	9
4	455, 461	3	3	2	8
5	432, 437, 464	1	5	5	11

35a List the type of crime police should focus on while patrolling the beat.

Student answers will vary. The following sample answers are based on the sample table above.

Beat: 416   Type of robbery to focus on: Individual
Beat: 419   Type of robbery to focus on: Individual
Beat: 421   Type of robbery to focus on: Residential
Beat: 412   Type of robbery to focus on: Business
Beat: 461   Type of robbery to focus on: Residential

MODULE 3: A WAR ON WEEDS

Exercise: Use GIS to map a leafy spurge infestation and compute its area

PART 1

2a Compare and contrast the geographic relief in the eastern and western parts of Bingham County. How does the map support your reasoning?

In the eastern part of the county I would expect to find hills or mountains, because the contours are close together and form tight circles. In the western part I would expect to find relatively flat or gently sloping land, because the contours are far apart and more linear.

2b Which is the Snake River and which is the Blackfoot?

The Snake River is on the west and the Blackfoot River is on the east.

3a Describe three distinct regions of Bingham County based on the distribution of the hydrologic features.

Region	Hydrologic characteristics
Mountain region in the east	A number of short rivers and creeks
Snake River Valley in the middle	The Snake River and a large number of irrigation canals and ditches
Plains region in the west	No hydrologic features

3b Rank the three regions according to where you would expect to find the most agricultural activity. Explain your reasoning.

The Snake River Valley region probably has the most agriculture because it would be relatively flat and has a lot of irrigation canals for watering the crops. The mountain region could have some agriculture, especially near the Blackfoot River. The western region probably has the least agriculture because of lack of water.

4a Does the annual precipitation map support the regions you described in question 3a and the ranks you gave them in question 3b? Explain.

Student answers will vary.

8a Describe the general pattern of public and private landownership in Bingham County.

Most of the privately owned land is found in a wide band along the Snake River and in the east. Large tracts of publicly owned land are found in the northwest, south of the Blackfoot River, and in scattered parcels in the east.

8b Name the three landowners who control most of the publicly owned land in Bingham County.

Bureau of Indian Affairs, Bureau of Land Management, State of Idaho

9a What observation can you make about the distribution of the leafy spurge points in your database relative to landownership?

Most of the leafy spurge points are on private land.

13a What geographic patterns do you see, if any, in the coverage density of leafy spurge?

Student answers will vary. They may observe no pattern, or they may observe a slight pattern where the infestations are heaviest near the Snake River and get lighter as one moves away from the river. They may notice that some infestations are clustered together, while others are single dots.

13b What observations can you make about the location of the leafy spurge points relative to features other than landownership?

The leafy spurge points tend to be near roads and in flat areas. There are no leafy spurge points in the western part of the county, and only one point in the eastern part.

16a Make three observations about the landscape you see in the image.

Student answers will vary but could include the following:
- The Snake and Blackfoot Rivers run across the corners of the image.
- There are a lot of circular agricultural fields.
- There is a northeast–southwest feature in the landscape that breaks up some of the fields.
- A major highway parallels the Snake River on the southern side.
- Most roads form a north–south and east–west grid pattern.
- Fields close to the Snake River are much smaller than those away from the river.
- The northwest and southwest corners of the image do not appear to have crop fields.

20a List three features you see on the 1-meter-resolution aerial photograph.

Student answers will vary but could include the following:
- Rows of the crops
- Farmer's house and farm buildings
- Roads

23a How many records are in the table?

12

23b What type of GPS receiver was used to collect the data?

GeoExplorer® 3

23c Which two fields contain information about the location of the weeds?

Northing, Easting

124

26a What patterns do you see, if any, in the leafy spurge coverage data?

Student answers will vary. They may notice that the infestations are heaviest in the center of the cluster and lightest at its edges.

30a What patterns do you see, if any, in the infestation size data?

Students should notice the infestations are nearly all large (1 to 5 acres). The smaller infestations are all on the western side of the cluster.

PART 2

21a What is the area of the leafy spurge infestation?

96,034.87 square meters (Student answers will vary but should be approximately 96,000 square meters.)

21b Convert the area to acres (1 acre = 4,047 square meters).

23.73 acres (Student answers will vary but should be approximately 23 to 24 acres.)

21c Write two reasons why the total area figure you came up with represents an estimate, and not the actual area.

1. The circles I drew did not have the exact radius measurement.
2. The area of the circles represents the average of a range of possible sizes.

25a What is the area of the farmer's fields?

318,497.81 square meters

25b What is the area of the fields in acres?

78.7 acres

25c What percentage of the field area is infested with leafy spurge?

Approximately 30 percent

28 Write a brief report to the farmer that answers his questions outlined in the scenario at the beginning of the exercise. Include your analysis results from part 2, steps 21 and 25.

Student answers will vary. An example answer is:
I found that the leafy spurge infestation in your field covered approximately 24 acres in total. The leafy spurge is currently taking up about 30 percent of your 78.7-acre field.

ASSESSMENTS

Average acres with LS	Radius of infestation in feet		
	Initial year (feet)	Projected growth after four years	Projected radius size after four years
0.05	26	60	86
0.55	87	60	147
3	204	60	264

3a Predicted total area of infestation in four years:

141,894.85 square meters or 35 acres (Student answers will vary slightly.)

3b Percentage of the total field area taken up by the infestation in four years:

Approximately 44.5 percent

MODULE 4: TRACKING WATER QUALITY
Exercise: Analyze Turtle River data to identify locations for fish habitat restoration

PART 1

4a	Name the two units in the watershed just south of the Turtle River.	Sandhill-Wilson and Goose
4b	Name the watershed unit that the Turtle River is in.	Turtle
4c	Which direction does the Turtle River flow?	Eastward and then north
6a	What types of aquifers affect the Turtle River?	Sand and gravel (glacial), sandstone, and sandstone and limestone
6b	Use the map and the information in step 5 to describe two ways you think the aquifers may affect the water quality of the Turtle River.	It is possible that the sandstone (glacial) aquifer would increase the salt concentration of the river water. The sandstone and limestone aquifer could make the water more basic (high pH).
8a	Which seasons are included in this data set?	Summer, fall, and winter
8b	What value and unit of measure are displayed for Dissolved Oxygen at Site #4 in the summer?	10.10 mg/l or 10.10 ppm
8c	What value and unit of measure are displayed for Conductivity at Site #5 in the winter?	523.5 microsiemens
8d	Based on the data, which sites have the coldest temperature and what is that temperature?	Sites 2 and 3
12a	Describe the distribution pattern of 3season_abc.shp.	All points are located approximately equidistant from each other and along the Turtle River.

17a Which of the sites meet the water depth criteria and what are their depths?

Site #	Water depth
1	3.8
2	5.51
3	4.2
4	3.1

PART 2

10a	Name the site(s) with the most acidic water measurement.	Sites 3 and 5 (7.84 in winter)
	Name the site(s) with the most basic water measurement.	Site 4 (8.56 in fall)
10b	Describe the general pattern of pH from summer to spring.	The pH decreases.
10c	Based on the information you learned about aquifers (step 5 of part 1), how would you expect the pH to change as the river flows eastward? Why?	Because the sandstone and limestone aquifer will increase the pH of the water, you can expect that the sites with the highest pH are sites 1 and 2 (closest to the mouth of the river and the sandstone and limestone aquifer).

10d Determine which three sites meet the pH requirements of the fish habitat restoration project (pH average closest to 7.0) and complete a table like the one below.

Site #	Su_ph	Fa_ph	Wi_ph	Sp_ph	Average pH
1	8.18	8.05	7.92	8.34	8.12
2	8.24	8.28	8.03	8.17	8.18
5	8.40	8.31	7.84	8.41	8.24

12a	What pattern, if any, exists in the dissolved oxygen levels from each site to the next?	There is no identifiable pattern.
12b	What trend, if any, exists in the dissolved oxygen levels from one season to the next?	Overall, the amount of dissolved oxygen is lowest in summer and highest in winter, increasing from summer to winter and decreasing from winter to summer. For most of the sites, winter is the season with the highest dissolved oxygen.

12c Explain the trend identified in 12b.

Gasses dissolve more easily in cold water. Therefore, it is not surprising that summer would have the lowest dissolved oxygen results and winter the highest.

12d Determine which sites meet the dissolved oxygen requirements of the fish habitat restoration project. Complete a table like the one below.

Site #	Summer	Fall	Winter	Spring
4	10.10	11.40	15.14	14.02
5	10.30	13.60	13.90	15.89

14a What pattern, if any, exists in the conductivity levels from each site to the next?

The conductivity increases as the river moves eastward.

14b Explain the high conductivity levels at sites 1 and 2.

The sandstone aquifer contributed a large amount of salt to the water, increasing its conductivity.

14c What trend, if any, exists in the conductivity levels from one season to the next?

There is no identifiable trend.

14d Determine three sites that meet the conductivity requirements of the fish habitat restoration project. Complete a table like the one below.

Site #	Su_conduct	Sp_conduct
3	1044.0	582.9
4	1045.0	588.0
5	408.0	605.9

15a Complete a table like the one below by placing an × next to each site number under the requirements it meets.

Site #	Water depth	pH	Dissolved oxygen	Conductivity
1	×	×		
2	×	×		
3	×			×
4	×		×	×
5		×	×	×

15b Based on your completed table, which site or sites would you recommend for fish habitat restoration?

Why?

Site 4

Both site 4 and site 5 meet three of the five requirements. However, site 5 is too shallow and will freeze during the winter, killing all the fish. Therefore, site 4 is the most suitable.

MODULE 5: INVESTIGATING POINT-SOURCE POLLUTION
Exercise: Map, query, and analyze neighborhood data to identify high-risk landfills

PART 1

8a	How long is the east–west axis?	9 kilometers
8b	How long is the north–south axis?	8.5 kilometers
9a	What is the predominant land use in the neighborhood?	Residential
9b	What, if any, land-use patterns are associated with Rivers?	Parks and recreational
9c	What, if any, land-use patterns are associated with Major Roads?	There is no single use that predominates, but most of East York's Commercial or Resource and Industrial land use is located on major roads.
12a	How many schools are there in the East York neighborhood?	119
15a	Which areas of East York are densely populated (more than 5,000 people per square kilometer)? Describe the location of specific areas by referring to major roads by name. Describe two more areas.	Answers will vary. Possible responses include: • The area in the center of the neighborhood bounded by Ellington Avenue E on the north, Don Mills Rd. on the west, and Don Valley Parkway on the east • The southern part of the neighborhood bounded by O'Connor Drive on the north and Broadview Ave. on the west • The west central part of the neighborhood bounded by Bayview Ave. on the east, St. Clair Ave. E on the south, and Ellington Ave. E on the north
15b	Give two possible reasons why some residential sections might have a higher population density than others.	Answers will vary. Possible responses include: • In neighborhoods where properties are larger, the population density will be lower, and conversely, where properties are smaller, the population density will be higher. • Areas that have multifamily housing such as apartment buildings or condominiums will have higher population densities.
15c	Is East York more densely populated in the northern half or the southern half of the neighborhood?	Southern half

PART 2

3a	How many landfills are located in East York?	28
3b	Are there more landfill sites in the northern half or the southern half of the neighborhood?	Southern half
3c	Based on your answer above (3b) and your answer to question 15c in part 1 about East York's population density, what generalization could you make about the difference between the northern and southern parts of the neighborhood?	Northern half: Lower population density, probably larger properties, only one landfill site Southern half: Higher population density, probably smaller properties or multifamily housing, 27 landfill sites
3d	Which category do you think is likely to present the greater danger, A3 or A5? Explain.	Answers will vary. Strict standards for landfill construction have been mandated in recent years, but were not in place when many of the older A5 sites were constructed and closed. Therefore the older sites probably represent a slightly greater danger.
4a	How many of East York's twenty-eight landfills are within 0.5 kilometer of a river?	21
5a	How many of East York's twenty-eight landfills are within 0.5 kilometer of a school?	17
5b	How many of East York's 119 schools are within 0.5 kilometer of a landfill?	16

PART 3

2a How many of East York's twenty-eight landfills are very close to rivers or schools?

11

5a Based on this analysis, how many of East York's twenty-eight landfills are designated high risk?

7

5b Is there any spatial pattern in the location of the selected sites? If so, describe it.

The seven sites are clustered in the southeastern corner of the neighborhood, generally along the path of the river.

5c Which river is in the greatest danger from these high-risk sites?

Massey Creek

5d Identify schools that are within 0.25 kilometer of any high-risk landfill.

Presteign Heights Public School, D. A. Morrison Elementary School, Canadian Martyrs Separate School (If students use the Measure Tool to answer this question, they might add Parkside Public School to the answer.)

MODULE 6: GETTING KIDS TO SCHOOL
Exercise: Use buffers to identify eligible school-bus riders

PART 1

1 Write three geographic questions that you could investigate based on the above scenario.

Student answers will vary but should include questions like the following:
- Where do the students live who may take the bus?
- Where do the students live who must walk to school?
- How many grades 1–3 students live more than 1,000 meters from school?
- Where is the area within (or beyond) 1,000 meters of the school?

3a Where do the students live in relation to the school? For example, do more students live near the school or far from it? Are they evenly spread out in all directions? Describe any patterns you see.

There is a large cluster of students surrounding the school, a substantial cluster to the southeast, and a small cluster scattered to the northwest.

16a, 19a Use the information to complete the table.

Student answers may vary slightly depending on the exact radius and location of their buffers.

Zone	Distance	Number of students
Zone 1	Within 1,000 meters	370
Zone 2	Between 1,000 and 1,600 meters	26
Zone 3	Beyond 1,600 meters	155
	Total students	551

PART 2

8a Compare the locations of the 1,600-meter buffer and 1,600-meter service area. Could the method you use to create distance zones affect your analysis results? Why or why not?

The two buffers mostly overlap, but the zone 2 (service area) buffer is shifted toward the north of the school. Analysis results could be somewhat different depending on which region is used, especially for students living at the northern and eastern edges of the buffers. Analysis results probably would not be drastically different, because in the large area south of the school that is included in the 1,600-meter circular buffer, there are no streets.

14a Refer back to the Busing Guidelines at the beginning of the exercise. Which grades are correct in the table below?

Zone 1	Distance from school: 0–1,000 meters							
Grades that ride the bus	**JK**	**SK**	1	2	3	4	5	6

17a What happened to the selected set of students on your map?

Some of the students were removed from the selected set.

17b How many kindergartners live in zone 1?

92

18a What are the correct grades for zone 2?

Zone 2	Distance from school: 1,000–1,600 meters							
Grades that ride the bus	**JK**	**SK**	**1**	**2**	**3**	4	5	6

21a Refer to question 18a and complete the following expression that will select the zone 2 students who may ride the bus. Choose from among the <, >, or = symbols.

([Grade] <= "3") or ([Grade] >= "JK")

22a How many students in zone 2 are eligible to ride the bus?

20

23a What are the correct grades for zone 3?

Zone 3	Distance from school: Beyond 1,600 meters							
Grades that ride the bus	**JK**	**SK**	**1**	**2**	**3**	**4**	**5**	**6**

25a How many students in zone 3 are eligible to ride the bus?

152

28a, Copy the data from the summary table into a
31a table like the one below.

Eligible bus riders by grade

Grade	Total number of students	Number of students eligible to ride the bus
JK	70	70
SK	68	68
1	65	23
2	69	23
3	71	23
4	65	16
5	69	18
6	74	23

33a How many eligible bus riders are there in all?

264

33b How many buses will be needed to transport them?

5 or 6

MODULE 7: PROTECTING THE COMMUNITY FOREST
Exercise: Map and query a tree inventory to locate hazardous trees

PART 1

4a	Describe what you see in the area enclosed by the semicircular drive.	There are trees along the street and along the driveway.
5a	How would you describe the area between the driveway and the lower property line?	Heavily wooded
19a	Identify two ways that tree 3 differs from the other two trees.	1. Tree 3 is a different species from the other two trees. 2. Tree 3 has a much larger diameter than the other two trees.

PART 2

3a	What is the ownership and health of tree 5?	The ownership is public and the health is fair.
3b	Of trees 2–10, how many have conflicts with overhead utility lines or adjacent trees?	7
5a	How many red oaks are there in this inventory?	15
6a	What do the yellow highlighted trees represent?	All of the red oak trees
6b	What is their distribution around campus?	There is a cluster of red oaks in the southeast section of the campus. Additional red oaks are along the northern boundary of the campus.
9a	What can you conclude about the age of the tree population on the Barrington Middle School property?	Mature
9b	List three questions about Barrington Middle School's tree population that you could answer and analyze with this data. If you have time, try to answer your questions.	Student answers will vary. They could ask questions requiring information in the attribute table that includes the species of trees, their height, health, and DBH. Students could also ask questions about trees that are in conflict with the school, sidewalk, utility lines, or each other. Some students may ask questions that involve a combination of any of the data attributes.
11a	What has changed on the map?	Many of the trees along the eastern boundary of the campus were selected.
11b	What do these changes mean?	All of the selected trees are in conflict with overhead utility lines.
12a	How did the map change?	There are more trees selected, especially between the south side of the school building and the driveway.
13a	How many of the middle school's 217 trees are hazardous?	40
13b	Is there any pattern to the location of the hazardous trees? If so, where are they?	Most of the hazardous trees are located along the street east of the school.

MODULE 8: SELECTING THE RIGHT LOCATION
Exercise: Perform site selection for a state wildlife area

5a	What features exist along the fence?	Two pieces of farm equipment and a pump house
5b	Based on the information in the map, what is the purpose of the pipe in the southwest section of the wildlife area?	The map displays a shed at the end of the pipe. It is possible that the pipe once served the shed as either a water line for irrigation or a sewage line.
5c	How could some of these features limit the area where the parking lot could go?	If the structures like the pump house, shed, and farm equipment cannot be moved, then the parking lot will have to be sited around them. The ditches will prevent cars from easily traveling in that section of the wildlife area. If CDOW wants to use preexisting roads to access the parking lot locations, then the lot will have to be sited near them and they exist only in some sections of the wildlife area.
8a	Why doesn't the aerial photograph display even though the theme's check box is checked?	In step 7, the maximum scale for the aerial photograph display was changed to 15,000. Therefore, it will not display at a scale of 16,000, even if the theme is turned on.
9a	The Yampa River theme was digitized from the aerial photograph. Describe the changes you see, if any, in the location of the Yampa River since 1969 (as shown on the USGS map).	In the northern section, the path of the river flow has changed dramatically. In the southern section, the river path has moved westward (much closer to the abandoned car and shed).
9b	Compare and contrast the geographic information about the wildlife area that is available from the USGS topographic map, the aerial photograph, and the point and line themes. Use specific features as examples of your observations.	Elevation contours and text information such as benchmarks are found only on the topographic map. The aerial photo shows more detail of the landscape; for example, you can see the ditches, fields, and details of the river course. Point and line features such as the pump house, shed, ditch points, fence, and corral are not visible on either the topographic map or the aerial photograph. Some GPS line features, like the ditches and the service road, are visible on the aerial photograph.
9c	Give an example of when you would use a topographic map and when you would use an aerial photograph.	Student answers will vary. An example of when to use a topographic map is when you need to evaluate elevation or slope (this data is not present in an aerial photograph). An example of when to use an aerial photograph is when you need to determine what land area is wooded or open fields (this data is not present in a topographic map).

17a	Why do you think CDOW has a required buffer of 300 feet around the river and why would they prefer a 500-foot buffer?	Student answers will vary. CDOW will want to protect the river from increased environmental damage caused by a nearby parking lot (for instance, increased sedimentation in the river, increased trash around the parking lot, noise).
18a	What general areas do you see, if any, that would be an appropriate site for parking lot option 3?	Student answers will vary but should exclude the areas covered by the buffers. Students should recognize that much of the SWA is eliminated from consideration.
19a	How could flood-zone data help you determine where parking lots should be located?	Student answers will vary. In general, parking lots should not be located in a flood zone. Frequent flooding would be hazardous to the people using the parking lot. The river would have increased sediment and the Colorado Division of Wildlife would likely spend more money fixing the damage caused by flooding.
19b	How have the possible sites changed with the addition of the flood-zone theme?	Student answers will vary. The flood zone further restricts the possible sites, especially in the southwestern region of the SWA.
20a	What differences do you notice between the locations of parking lot options 1 and 2?	Students may observe that option 1 is in the flood zone; it is farther south, it is near the intersection between CR14 and CR14F, and it is near the service road. Option 2 is along a straight section of CR14 and at the intersection of CR14 and Elk Lane; it is farther north, on a hill, and within a fenced area.
23a	List the reasons why you chose this particular location for your site.	Student answers will vary.
30a	Choose the location that best meets the criteria.	Student answers will vary.
30b	Explain the advantages of your chosen site over the other two site locations.	Student answers will vary. Students should explain reasons for preferring or rejecting each of the three options.

Community Geography: GIS in Action Teacher's Guide

Book design, illustration, production, and copyediting by Michael J. Hyatt

Cover design and production by Wendy Brown

Printing coordination by Cliff Crabbe